MCQs in Pharmaceutical Science and Technology

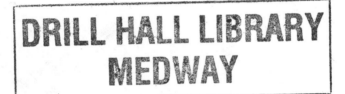

MCQs in Pharmaceutical Science and Technology

Edited by
Sanjay Garg

BPharm, MPharm, PhD, MMgt

Associate Professor and Deputy Head,
School of Pharmacy
University of Auckland
Auckland, New Zealand

London • Chicago   **Pharmaceutical Press**

Published by the Pharmaceutical Press

1 Lambeth High Street, London SE1 7JN, UK
1559 St Paul Avenue, Gurnee, IL 60031, USA

© Pharmaceutical Press 2011

(**PP**) is a trade mark of Pharmaceutical Press
Pharmaceutical Press is the publishing division of the
Royal Pharmaceutical Society of Great Britain

First published 2011

Typeset by Thomson Digital, Noida, India
Printed in Great Britain by TJ International, Padstow, Cornwall

ISBN 978 0 85369 913 2

A catalogue record for this book is available from the British Library.

Dedication

To the world's most wonderful mom, Bimla Devi, who taught me the art of pharmaceutical science. Her cooking always inspired me to be a product development scientist and I have always used 2 versus 20 cups of tea as an example to explain scale-up issues. Her desire to help others motivated me to write this book.

Contents

Foreword

Pharmacy education has moved over the past few decades from being medicine-centred to patient-centred. Pharmacists are no longer just dispensers, or perhaps not even dispensers, but rather are experts on medicines who can help in the optimal choice of medicines and assist patients to make the best use of their medicines.

When major shifts in thinking and function occur, there is the danger that traditional knowledge and skills are downplayed or even seen as unnecessary – the proverbial 'throwing the baby out with the bathwater'. But can a pharmacist be an expert on medicines if he or she is knowledgeable only in pharmacotherapeutics?

To be truly a medicines expert, the pharmacist needs to be an expert in three major areas: (i) on the product, i.e. both the drug and the formulation, (ii) on the patient – from biological, behavioural and social perspectives, and (iii) on the effects of the product on the patient and the patient on the product. This is all set within a society's healthcare system with its ethicolegal, professional, consumerist and economic frameworks. In terms of a simple mnemonic, the pharmacist has to be an expert of the P (product), the P (patient) and the P-in-the-P.

Being an expert on the product or medicine requires knowledge of the active pharmaceutical ingredient, the drug, and also an understanding of the dosage form. What goes into the formulation and why and how those excipients might influence the behaviour of the dose form in the bottle and in the body, and the patient's response (e.g. a lactose-intolerant patient), should be understood by the medicines expert. Consequently, pharmacy students have been taught for many years about the ingredients and their roles in manufacture and stability of medicines and how they affect behaviour of dose forms (e.g. disintegration, release behaviour) in the patient. Equally as important, students have been taught physical pharmacy, the science that underpins much of the behaviour of dose forms, thereby providing a basis for reasoned answers to 'what if?' questions.

New understanding of the effects of excipients suggests even greater complexity than previously envisaged when we thought these ingredients innocuous. We now understand that so-called 'inactive' ingredients might affect transporters and other absorption processes and thereby modify bio-availability. The effects of food and types of food on absorption continue to be elucidated. For example, alterations in bioavailability due to effects on gut-wall metabolism or shifts from absorption via the portal system to the lymphatics, with subsequent changes in the side-effect profile, challenge the expert on medicines who wishes to optimise the use of a medicine in a particular patient.

The past 30 years have seen active research on novel drug-delivery systems with the aim of delivering drugs with greater finesse – the right drug, at the right time, at the right rate, to the target site. Although ideal delivery profiles or target specificity have not been achieved, important advances have been made. There are currently more than 2500 controlled-/sustained-release products in development, across all therapeutic classes, with most being in phase 2 trials. These novel technologies will present new challenges in the clinic, and pharmacists will be called upon for their expert advice, not only on patient-specific issues related to a particular technology but also on economic issues and whether the novel technology is justified.

Although the majority of medicines are given orally, and this is likely to continue for convenience reasons, huge strides have been made in the development of sophisticated formulations for non-oral routes: injectables, transdermals, pulmonary aerosols, and so on. Although there have been some disappointments (e.g. pulmonary insulin), the field has made some major advances in providing products with improved delivery profiles and fewer side-effects for patients. However, these novel medicines bring their challenges to the medicines expert, who, for example, needs to understand what influences the absorption rate of a drug through the skin and how this might be influenced by the skin site, the cleansers used and the health status of the skin.

Testing the knowledge of the future experts on medicines is an important task for those involved in the education of pharmacists. We are not interested simply in testing essential factual knowledge (although some factual knowledge is essential); we also want to ensure that students have developed the ability to use or apply their knowledge to provide patient-specific advice. That is, we want to test so-called deep learning. Some argue that multiple choice questions (MCQs) can be used only to test surface learning, but this is not my view. Discussions with educationalists whose research field is 'assessment' have convinced me that well-designed MCQs can be used to test deep learning. The advantage of MCQs over free-form answers is that they are easily marked, an important consideration given constrained resources and heavy demands on teachers' time; the disadvantage of MCQs is that good questions take a long time to compose.

It is therefore good that the contributors have given their time to this useful book of MCQs. I hope it will be a valuable resource for those involved in the education of the experts on medicines. It certainly covers the important areas discussed above.

Professor Ian G Tucker
Dean
New Zealand's National School of Pharmacy
University of Otago, New Zealand

Preface

This book, edited by Sanjay Garg, with contributions from a range of distinguished pharmaceutics educators, provides a much-needed examination and revision guide for pharmacy students in their study of pharmaceutics. It is particularly timely given the increasing trend to use multiple choice questions (MCQs) as a significant part of the assessment procedures in pharmacy degree courses. The book can be viewed on two levels: first as a self-test to guide students during their learning, giving them an insight into their progress, and second as a revision primer for degree examinations. Most pharmaceutics educators will recognise the content as core to any pharmacy course and can thus be confident that it will be of significant value to their students.

The book contains 600 MCQs divided into six tests, progressing logically from basic science through to clinical considerations in dosage form design, advanced dosage forms and, finally, regulatory and industrial aspects. Indeed, this also approximates the sequence for the development of a pharmaceutical product incorporating an advanced delivery system. Thus, in Test 1, the student is presented with MCQs relating to such basic aspects of pharmaceutics as solubility and dissolution, kinetics and stability, ionic equilibria and rheology. Arguably, if a student can master these and related topics, then the rest is easy! Importantly, clinical aspects relating to dosage form design, and particularly the design of oral dosage forms, are covered early in the book, in Test 2, with an emphasis on biopharmaceutics and pharmacokinetics. Underlying aspects of dosage form design are dealt with in a test on particulates, leading on to a range of MCQs covering basic and advanced drug-delivery systems. The final test challenges the student on their knowledge of regulatory issues presented by dosage form design, together with aspects of intellectual property. In effect, these questions relate quite well to the chemistry and pharmacy sections of a regulatory dossier.

Throughout the book, the format of MCQs used is simple and straightforward. This is commendable, since the object is to help the student in their learning, not to confuse or trick them in any way. Each test is provided with succinct but sufficiently detailed answers, probably more detail than a student might actually give in answering a question under real examination conditions but, importantly, enough information to help them learn and understand the topic being studied. I recognise many of the questions as similar to those used in the examination of pharmaceutics learning in my own school, and I am sure colleagues will similarly recognise much of the content.

Overall, the subject matter of the MCQs will be relevant to most schools of pharmacy that teach pharmaceutics as the core science that is unique to pharmacy, since the design of dosage forms and advanced drug-delivery systems requires a knowledge of chemistry (physical, polymer, organic, biochemical and analytical) integrated with a sound understanding of biology and clinical pharmacy. Quite a challenge for any student and one that Sanjay and his colleagues, through this excellent book, will help them to meet.

Professor David Woolfson
Head
School of Pharmacy
Queen's University Belfast
Northern Ireland, UK

Acknowledgements

This is my first edited book and a very special project. During writing, I realised the value of efforts of some very special people, which helped me in completing it in the stipulated time.

My heartfelt thanks to the contributors, Alison Haywood, Beverley Glass, David S Jones, Omathanu Perumal, Paul Ho, Roop K Khar and Therese Kairuz, for working with me on their tests and the good discussion we had during writing. Their support has helped me in giving the book a true international platform. Knowing the busy schedule of academics, it was pleasing to see them according a high priority to the book. I express special thanks to Professor Ian Tucker and Professor David Woolfson for taking the time to write the foreword and the preface at short notice.

I would like to acknowledge the constant support and encouragement of Professor John Shaw, Head, School of Pharmacy, and the University of Auckland for sabbatical. My sincere thanks are due to Darren, Mandy and Meghna for wonderful help with proofreading. Darren and Meghna read the individual sections from a student's perspective and provided highly valuable suggestions. The pharmacy students from Auckland, who provided valuable feedback to MCQs during examination, deserve a special mention. Thanks are due to Rebecca Perry and Linda Paulus from the Pharmaceutical Press, who have been highly patient and supportive of this publication.

Special thanks to my kids, Shubham and Astha, with whom I shared my feelings and occasional frustrations. I am short of words in expressing my thanks to my love and best friend Alka, for her unconditional support and understanding. Her input from a clinical pharmacy perspective has been of great value in giving a direction to the book. Thanks to all other friends and family for their support, keen interest in the book and motivation.

About the editor

Sanjay Garg studied pharmacy at Delhi University in the form of bachelor's and master's degrees in pharmaceutics. His master's research project on aerosol formulations led to a research-driven career, completing a PhD at the National Institute of Immunology, India, and postdoctoral research at Rush University, Chicago, USA. He joined the Program for Topical Prevention of Conception and Disease (TOPCAD), USA, as assistant professor in 1996. During his tenure with TOPCAD, Sanjay developed an interest in regulatory affairs and project management. In the period 1998–2003, he played a key role in developing new master's level papers in pharmaceutical sciences and project management at the National Institute of Pharmaceutical Education and Research (NIPER), India. Since 2003, Sanjay has been engaged as deputy head at the School of Pharmacy, University of Auckland, New Zealand, where he is actively involved with undergraduate and postgraduate programmes. At Auckland, he established AnQual, a GLP-compliant analytical facility, and works in close collaboration with research groups, foundations and pharmaceutical industries from several countries. His research interests are in the development of conventional and novel formulations, microbicides for HIV prevention, tumour targeting, nanotechnology and drug policy.

Sanjay has over 100 publications, 20 patents and a number of awards to his credit. He is frequently invited to international conferences. He is an editorial board member for pharmaceutical technology, bioanalysis, and clinical and experimental ophthalmology, and he acts as a reviewer for several other journals. Throughout his career, Sanjay has enjoyed interacting with pharmacy students. He chairs a students' affairs committee and acts as a board member of the International Society for Pharmaceutical Engineering (ISPE), Australasia affiliate. Sanjay works in collaboration with academicians from the USA, Brazil, Singapore, Malaysia, Australia, India and China. The result of this networking is reflected in this book, which is based on contributions from friends in six countries.

This book has been a special project for Sanjay and is an outcome of his keen desire to help pharmacy students and staff from all over the world. Any feedback and suggestions about the book can be sent to him by email at s.garg@auckland.ac.nz

Contributors

Alison Haywood
Senior Lecturer, School of Pharmacy, Griffith University, Gold Coast,
Australia

Beverley D Glass
Professor and Head, School of Pharmacy and Molecular Sciences,
James Cook University, Townsville, Australia

David S Jones
Professor, Chair of Biomaterial Science, School of Pharmacy,
Queen's University Belfast, UK

Omathanu P Perumal
Associate Professor, College of Pharmacy, South Dakota State University,
Brookings, SD, USA

Paul Chi-Lui Ho
Associate Professor, Department of Pharmacy, National University of
Singapore, Singapore

Roop K Khar
Professor of Pharmaceutics, Faculty of Pharmacy, Hamdard University,
New Delhi, India

Sanjay Garg
Associate Professor and Deputy Head, School of Pharmacy, University of
Auckland, Auckland, New Zealand

Therese Kairuz
Senior Lecturer, School of Pharmacy, University of Queensland,
Woolloongabba, Australia

Introduction

Pharmaceutics is generally considered to be associated with the design and manufacturing of dosage forms. In my experience, pharmaceutical inputs are required from the drug discovery stage to eventual administration of the dosage form to the patient. For example, in new drug discovery programmes, preformulation experiments normally provide the necessary component of the go/no-go matrix for selecting the right lead candidates. At the other extreme, crushing of a tablet and dispersing into a syrup base for dispensing to a paediatric patient or extemporaneous compounding also involves principles of pharmaceutics. The depth and focus of pharmaceutics vary a lot in different countries. In countries with a strong pharmaceutical industry, students have to learn a great deal about industrial pharmacy, processing, scale-up, and so on. Countries with a stronger focus on clinical pharmacy tend to take pharmaceutics differently, with less emphasis on industrial and more on clinical pharmaceutics. Practices such as extemporaneous compounding are relatively less common in some countries, while in others they tend to be an integral part of the pharmaceutics curriculum and practice.

To design and develop a delivery system with the desired target product profile, it is important to have good understanding of good practice (GXP), i.e. good laboratory practices (GLP), good manufacturing practices (GMP) and good clinical practices (GCP). Regulatory procedures and expectations govern the entire process of drug development. Modern concepts in drug manufacturing such as process analytical technologies (PAT) and quality by design (QBD) have added another dimension that is heavily pharmaceutics-oriented.

This book has been divided into six tests. We have tried to cover topics taught in most pharmacy schools around the globe. The sequence of topics and subdivision into different tests is based on gradual progression in the curriculum from year 1 to year 4. Test 1 starts with the basic principles of physical pharmaceutics, addressing the parameters required to be assessed for designing a dosage form. The test ends with questions on preformulation, where all these principles are applied to gather information about a drug molecule. Test 2 focuses on oral drug delivery, gastrointestinal characteristics and pharmacokinetics, and introduces the concept of controlled drug delivery. Test 3 is based on particle science and the design of solid dosage forms. Solid-state properties and unit operations such as size reduction, mixing, granulation and drying affect the processing of these dosage forms. The test involves a set of questions on pharmaceutical calculations, which is usually a critical component and requires serious efforts by students. Test 4 leads on to conventional dosage forms, i.e. liquids, semisolids and solids, covering their design, excipients, formulation, processing and evaluation. The modifications to novel delivery systems for various routes of administration are covered in Test 5. This test also provides a set of questions on intellectual property management, which is closely associated with novel delivery systems. Test 6 is much broader in scope, addressing topics that are supportive in nature, i.e. microbiology, extemporaneous compounding, regulatory affairs and packaging. Quality assurance and control, good laboratory practices, good manufacturing practices and good clinical practices are integral functions required to take a drug from bench to patient and with which pharmacists are involved in various ways. It is very difficult to cover the entire scope of pharmaceutics in one publication; this book is an attempt to address some of it.

In my lifetime, I have seen pharmacy classes with 18–300 students. Larger classes exert tremendous pressure on the assessment system, and that is the reason for the increasing popularity of MCQs. Properly designed questions are an extremely valuable assessment tool. This book has been designed with pharmacy undergraduate students in mind, keeping an international perspective. It will also cater to postgraduate students and will be of great value in entrance or pharmacy board examinations. Although it will serve as a good resource for teachers, it also has great potential as a revision tool for students. The style of the questions is mixed, with varying levels of difficulty. The questions have been reviewed extensively and validated. Emphasis has been given to case studies, wherever possible. Students are advised to understand their regular text before attempting to practise the questions in this book. They should try solving the questions without looking at the answers, followed by checking the answers and reading the associated text in the next stage. The text provided with the answers is very brief and not a replacement of regular textbooks; it has to be taken as supplementary only.

My contributor friends and I have made every attempt to check the correctness of the questions and answers. In drafting this book, we have referred to well-established pharmaceutical textbooks and pharmacopoeias. However, I will sincerely appreciate any corrections, feedback and suggestions by email (s.garg@auckland.ac.nz).

Test 1

Physical pharmaceutics

Paul Chi-Lui Ho, David S Jones, Sanjay Garg

and Roop K Khar

Design of dosage forms

Dissolution and solubility

Solution properties, disperse systems

Rheology

Ionic equilibria, buffers

Surface and interfacial phenomena

Kinetics and stability

Preformulation

Introduction

With advances in automated synthesis, combinatorial chemistry and high-throughput screening, a vast number of poorly water-soluble but bioactive drugs have been introduced in the pharmaceutical development pipeline. The poor pharmaceutical and biopharmaceutical properties of these drugs prevent many of them from reaching the market. It has been estimated that up to 60% of the total new chemical entities from the pharmaceutical chemical laboratories are water-insoluble, and over 40% of marketed drugs are poorly water-soluble. Based on the biopharmaceutics classification system (BCS), many drugs either in the development pipeline or in the market are categorised as BCS class II compounds, i.e. these drugs exhibit low solubility but reasonable membrane permeability, and the rate-limiting process in their absorption is in the dissolution steps. In this regard, formulation becomes an ever important subject to make these drugs 'soluble' and subsequently 'bioavailable' clinically. Physical pharmaceutics is the science of formulation or dosage form design that turns a drug into a medication that could be effectively administered to patients in order to exert its desired therapeutic effects. It is a discipline that encompasses both physical and chemical sciences. Physical pharmaceutics focuses on, but is not limited to, solubility or solubilisation and stability issues. It also covers solutions and their properties; rheology; ionic equilibria and buffers; surface and interfacial phenomena; and preformulation. These topics appear to be diverse in nature, but they can be interrelated with each other, and each of them could be an important factor in many pharmaceutical processes. For example, surface and interfacial phenomena play an important role in the processing of a wide variety of formulations. Starch is a commonly used excipient in tablet and capsule formulations. The surface tensions of starches from different plant sources vary, thus influencing the spreading behaviour of the binder over the substrate particle. High interfacial tension of the polymeric binder will result in formation of thick and hard layers of the binding film with poor moisture penetrability. The rate-limiting step in tablet disintegration is the penetration of dissolution media through the pores of the tablet, and the surface tension or adhesion tension in this circumstance is the driving force for liquid penetration into solid dosage forms. In dispersed systems such as emulsions, the interfacial tension between the dispersed and continuous phase determines the physical stability of the emulsion. It is not easy for students to grasp the concepts and understand these diversified but tightly integrated topics in physical pharmaceutics. These topics also pose different levels of difficulty in understanding them. It will be helpful for students to have a holistic overview and understanding of the subjects. The MCQs in this session will serve this purpose. It is advisable for students to attempt the questions before they look at the answers and their explanations. In modern pharmaceutical development, there are many novel dosage forms in the pipeline or in the market. However, fundamental knowledge is still required in order to appreciate and evaluate these products and, more importantly, to further develop many more new products. We hope the MCQs in this test will help students grasp the concepts of this important field in the pharmaceutical sciences.

Questions

Q1　The main objective(s) of dosage form design is (are) to ensure:

1　❏ reproducible product quality
2　❏ user acceptability
3　❏ uniformity in drug response

A ❏ 1 and 2
B ❏ 2 and 3
C ❏ 1 and 3
D ❏ 1, 2 and 3
E ❏ 1

Q2 The base form of prednisolone is most suitable to be formulated as:

1 ❏ a tablet for oral administration
2 ❏ an eye drop
3 ❏ an enema
4 ❏ a slowly absorbed intramuscular suspension injection
A ❏ 1 and 2
B ❏ 2 and 3
C ❏ 1 and 3
D ❏ 1 and 4
E ❏ 2

Q3 The sodium phosphate salt of prednisolone is most suitable to be formulated as:

1 ❏ a tablet for oral administration
2 ❏ an eye drop
3 ❏ an enema
4 ❏ a slowly absorbed intramuscular suspension injection
A ❏ 1 and 2
B ❏ 2 and 3
C ❏ 1 and 3
D ❏ 1 and 4
E ❏ 3

Q4 The time for onset of action for different dosage forms is in the following order:

A ❏ intramuscular injection > enteric coated tablet > tablet > intravenous injection
B ❏ intravenous injection > intramuscular injection > enteric coated tablet > tablet
C ❏ enteric coated tablet > tablet > intramuscular injection > intravenous injection
D ❏ tablet > enteric coated tablet > intramuscular injection > intravenous injection
E ❏ enteric coated tablet > intramuscular injection > tablet > intravenous injection

Q5 Which of the following statements about particle size reduction is incorrect?

A ❑ particle size reduction results in an increase in the surface area
B ❑ particles with smaller sizes generally dissolve at a faster rate
C ❑ particle size reduction can always improve drug bioavailability
D ❑ the bioavailability of spironolactone can be improved by particle size reduction
E ❑ optimising the particle size of drugs is a common practice in the pharmaceutical industry

Q6 Which of the following statements about the crystal properties of drugs that exhibit polymorphism is incorrect?

A ❑ the different crystalline forms vary in physical properties such as dissolution and solid-state stability
B ❑ a drug can exist in more than one crystalline form
C ❑ crystalline form transition can occur during milling
D ❑ different crystalline forms vary in hardness, shape and size
E ❑ it is usually difficult to identify and separate polymorphic forms after the drug is incorporated in a formulation

Q7 Surfactants can be used as:

1 ❑ emulsifying agents
2 ❑ solubilising agents
3 ❑ wetting agents
4 ❑ antiseptics
A ❑ 1 and 2
B ❑ 2 and 3
C ❑ 1, 2 and 3
D ❑ 1, 2 and 4
E ❑ 1, 2, 3 and 4

Q8 Which of the following statements is (are) correct?

1 ❑ the amorphous form of a drug is always more soluble than the corresponding crystalline form
2 ❑ generally, the anhydrous form of the drug dissolves more rapidly in water than the hydrous form
3 ❑ solubility of a weak acid can be increased by adding a conjugate base
A ❑ 1
B ❑ 1 and 2
C ❑ 3

D ❏ 2 and 3
E ❏ 1, 2 and 3

Q9 Erythromycin undergoes acid-catalysed hydrolysis in gastric acid. Its stability in gastric acid can be improved by:

1 ❏ formulating it in enteric dosage form
2 ❏ forming erythromycin estolate
3 ❏ administering with meals

A ❏ 1
B ❏ 2
C ❏ 3
D ❏ 1 and 2
E ❏ 1, 2 and 3

Q10 Lozenges are tablets that:

A ❏ should be chewed before swallowing
B ❏ are intended to be dissolved slowly in mouth
C ❏ are film-coated for prolonged action of medicaments
D ❏ are sugar-coated to mask the unpleasant taste
E ❏ are formulated with disintegrants in the formulation

Q11 In order for dissolution to occur spontaneously, the change in free energy must be:

A ❏ positive first, and then zero at equilibrium
B ❏ negative first, and then zero at equilibrium
C ❏ positive first, and then negative at equilibrium
D ❏ negative first, and then positive at equilibrium
E ❏ negative in the beginning and at equilibrium

Q12 According to Fick's law of diffusion, the rate of diffusion is directly proportional to the:

1 ❏ concentration difference between the two sides of the diffusion layer
2 ❏ difference in concentration of solution at the solid surface (C_1) and the bulk of the solution (C_2)
3 ❏ difference in concentration of the saturated solution (C_s) in contact with the solid at equilibrium and the bulk of the solution (C)

A ❏ 1
B ❏ 1 and 2
C ❏ 1 and 3

D ❑ 2 and 3
E ❑ 1, 2 and 3

Q13 The dissolution rate of solids in liquids increases with increase in:

1 ❑ temperature
2 ❑ viscosity of the dissolution medium
3 ❑ particle surface area
4 ❑ diffusion coefficient
5 ❑ diffusion layer thickness
 A ❑ 1, 2 and 3
 B ❑ 2, 3 and 4
 C ❑ 3, 4 and 5
 D ❑ 1, 3 and 4
 E ❑ 2, 4 and 5

Q14 Which of the following factors does not affect the saturation concentration of a solid?

A ❑ temperature
B ❑ nature of the dissolution medium
C ❑ size of the solid particles
D ❑ crystalline form of the solid
E ❑ intrinsic solubility of the drug

Q15 The solubility of an electrolyte is higher in a medium with:

A ❑ a high dielectric constant
B ❑ a low dielectric constant
C ❑ a high proportion of alcohol
D ❑ A and C
E ❑ B and C

Q16 Mercuric iodide (HgI_2) is more soluble in a solution of potassium iodide because of:

A ❑ the common ion effect
B ❑ change in pH
C ❑ formation of a soluble complex
D ❑ A and B
E ❑ B and C

Q17 Solubility can be expressed as

A ❑ molarity
B ❑ molality

C ❏ mole fraction
D ❏ A and B
E ❏ A, B and C

Q18 Which of the following statements on the solubility of solids in liquids is (are) correct?

A ❏ the solubility of the solid usually increases with increasing temperature
B ❏ the solubility of some solids can decrease with increasing temperature
C ❏ the solubility of some solids increases with increasing temperature, but above a certain temperature the solubility decreases with increasing temperature
D ❏ A and B
E ❏ A, B and C

Q19 The objective(s) of performing dissolution tests is (are) to:

1 ❏ determine the solubility of the drug
2 ❏ predict the absorption pattern in vivo
3 ❏ serve as a quality control procedure
A ❏ 1
B ❏ 2
C ❏ 1 and 2
D ❏ 2 and 3
E ❏ 1, 2 and 3

Q20 Surfactants can function as solubilising agents for certain poorly soluble drugs because they can:

A ❏ form more soluble ionised salts with the drugs
B ❏ form more soluble complexes with the drugs
C ❏ alter the dielectric constant of the medium
D ❏ increase the viscosity of the medium
E ❏ alter the pH of the medium

Q21 The following properties are true concerning weak acids or weak bases:

A ❏ weak acids are essentially fully dissociated at pH 7.0
B ❏ weak acids are more soluble in solutions of low pH than in solutions of high pH
C ❏ the % ionisation of weak acids increases as the solution pH is increased

D ❏ the % ionisation of weak bases increases as the solution pH is increased

E ❏ A and C

Q22 Which of the following statements are true concerning amphoteric drug molecules?

A ❏ the oil/water partition coefficient increases as the pH is increased

B ❏ the solubility increases as the pH is increased

C ❏ the solubility increases as the pH is decreased

D ❏ the solubilities of different isomers of the same compound are always identical

E ❏ the presence of electron-withdrawing groups, e.g. Cl, on the drug molecule may increase the solubility

Q23 Solutions of proteins, e.g. albumin, exhibit the following properties:

A ❏ increased viscosity as the concentration of protein in solution decreases

B ❏ greatest solubility at the isoelectric point

C ❏ decreased solubility in the presence of dissolved electrolytes

D ❏ increased solubility in the presence of dissolved electrolytes

E ❏ increased osmotic pressure as the concentration of dissolved protein is decreased

Q24 In concentrated aqueous drug solutions:

A ❏ the concentration of dissolved drug is equivalent to the activity of the solution

B ❏ ion pairs may exist

C ❏ the chemical potential describes the electrical conductivity of the solution

D ❏ the solution will be hypertonic

E ❏ B and D

Q25 Which of the following statements is true for polyprotic drugs?

A ❏ the percentage ionisation is independent of the pH of the solution

B ❏ at least two pK_as exist

C ❏ only one pK_a exists

D ❏ the pK_a values must be within 1 pH unit of each other

E ❏ under physiological conditions the molecule will be completely ionised

Q26 The diffusion of drugs of molecular weight <700 in solution is dependent on:

A ❏ temperature
B ❏ viscosity
C ❏ molecular weight
D ❏ nature of the solvent
E ❏ temperature, viscosity and nature of the solvent

Q27 The following statements are true concerning solutions of non-ionic surfactants:

A ❏ increasing the concentration of surfactant will alter the pH of the formulation
B ❏ increasing the concentration of surfactant will lower the surface tension of the solution
C ❏ non-ionic surfactants may increase the solubility of drugs in solution
D ❏ A and B
E ❏ B and C

Q28 The presence of drugs in aqueous solution will result in which of the following:

A ❏ a decrease in the boiling point
B ❏ an increase in the freezing point
C ❏ an increase in the osmotic pressure
D ❏ a decrease in the osmotic pressure
E ❏ no effect on the osmotic pressure

Q29 The following factors may lower the solubility of basic drugs in solution:

A ❏ the presence of magnesium ions
B ❏ the presence of certain buffer components, e.g. phosphates
C ❏ the presence of anionic polymers, e.g. sodium carboxymethylcellulose
D ❏ the presence of suspended solids, e.g. kaolin
E ❏ B, C and D

Q30 Which of the following factors will enhance the stability of drug suspensions?

A ❏ increasing the pH of the suspension
B ❏ increasing the particle size of the dispersed drug

C ❑ decreasing the viscosity of the suspension

D ❑ lowering the ionic content of the suspension

E ❑ reducing the particle size and increasing the viscosity of the suspension

Q31 Which of the following statements is true regarding emulsions?

A ❑ oil-in-water emulsions are more hydrophobic than water-in-oil emulsions

B ❑ the stability of emulsions is partly controlled by the viscosity of the dispersed phase

C ❑ the effect of surfactants on the stability of emulsions is due exclusively to their ability to lower the interfacial tension between the two phases

D ❑ the stability of emulsions is partly controlled by the viscosity of the continuous phase

E ❑ water-in-oil emulsions can be readily diluted with water

Q32 Which of the following statements is true of oil-in-water emulsions?

A ❑ oil-in-water emulsions are inherently more stable than water-in-oil emulsions

B ❑ oil-in-water emulsions are electrically conductive

C ❑ oil-in-water emulsions are electrically non-conductive

D ❑ oil-in-water emulsions can never be injected intravenously

E ❑ the stability of oil-in-water emulsions is unaffected by changes in the viscosity of the external phase

Q33 Which of the following statements is true regarding controlled flocculation?

A ❑ controlled flocculation is the first step in suspension instability

B ❑ controlled flocculation is employed to ensure that the interaction between suspended drug particles occurs at the primary minimum

C ❑ suspensions in which controlled flocculation occurs result in a small sedimentation volume

D ❑ the sedimentation rate of flocculated systems is faster than for deflocculated systems

❑ the sedimentation rate of flocculated systems is slower than for deflocculated systems

Q34   Surfactants assist in the stabilisation of emulsions by:

A   ❏ increasing the viscosity of the external phase
B   ❏ decreasing the viscosity of the internal phase
C   ❏ forming an elastic film at the interface between the internal and external phases
D   ❏ increasing the solubility of the internal phase within the external phase
E   ❏ minimising the size of the dispersed phase

Q35   Emulsions may be stabilised by the use of particulate systems, e.g. kaolin, and the presence of macromolecules. The mechanism of this enhanced stability is due to:

A   ❏ adsorption of the solid particles at the interface between the internal and external phases, hence forming a protective film
B   ❏ enhancement of the viscosity of the external phase due to the presence of dispersed solid particles, e.g. kaolin, bentonite
C   ❏ enhancement of the viscosity of the external phase due to dissolved macromolecules
D   ❏ stabilisation of the internal phase due to multilayer adsorption of macromolecules on to the droplets of the internal phase
E   ❏ A, C and D

Q36   Which of the following statements is true concerning the zeta potential?

A   ❏ the zeta potential is the potential at the surface of the dispersed phase
B   ❏ the zeta potential directly contributes to the primary maximum, i.e. the repulsion barrier that resists intimate particulate contact
C   ❏ the magnitude of the zeta potential is not affected by the presence of electrolytes
D   ❏ the magnitude of the zeta potential may be dramatically reduced by the presence of non-ionic surfactants
E   ❏ the zeta potential has no effect on emulsion stability

Q37   Which of the following statements is true regarding particle sedimentation?

A   ❏ the rate of sedimentation of particles is enhanced by increasing particle size
B   ❏ increasing solution viscosity increases the rate of sedimentation of particles

C ❏ the rate of sedimentation of particles is increased by decreasing the density of the suspended phase

D ❏ particle shape has no effect on the subsequent rate of sedimentation

E ❏ the rate of sedimentation can always be controlled by adjusting the pH of the suspension

Q38 A flocculated suspension is associated with:

A ❏ a high zeta potential

B ❏ a small sedimentation volume and cloudy supernatant

C ❏ a large sedimentation volume and clear supernatant

D ❏ slow sedimentation

E ❏ difficult redispersion on shaking

Q39 An emulsion is a system that:

A ❏ contains only one phase

B ❏ is inherently thermodynamically unstable

C ❏ cannot flocculate

D ❏ can only use a surfactant as an emulsifying agent

E ❏ always requires a mixture of emulsifying agents

Q40 Lyophilic colloids:

A ❏ are water-insoluble

B ❏ may be employed to increase the viscosity of pharmaceutical preparations

C ❏ always have molecular weights below 5000

D ❏ are not useful in enhancing the stability of pharmaceutical suspensions

E ❏ are obtained from natural sources only

Q41 Which of the following statements is true concerning Newtonian flow?

A ❏ the dynamic viscosity may be defined as the ratio of the rate of shear to the shearing stress

B ❏ water, propylene glycol and polyethylene glycol are examples of Newtonian liquids

C ❏ the viscosities of Newtonian liquids are temperature-independent

D ❏ the flow behaviour of Newtonian liquids is affected by pH

E ❏ ionic concentrations of the dissolved solids have a direct impact on the behaviour of Newtonian liquids

Q42 Which of the following options is correct regarding pseudo-plastic flow?

   A   ❏ viscosity is shear-stress-independent
   B   ❏ viscosity is rate-of-shear-independent
   C   ❏ viscosity is temperature-dependent
   D   ❏ creams frequently exhibit pseudo-plastic flow
   E   ❏ C and D

Q43 Which of the following option is correct regarding the Mark–Houwink equation?

   A   ❏ the Mark–Houwink equation may be used to estimate the molecular weight of polymers
   B   ❏ the Mark–Houwink equation may be used to estimate the dynamic viscosity of creams
   C   ❏ the Mark–Houwink equation may be used to determine the specific viscosity of dilute polymer solutions
   D   ❏ all of the above
   E   ❏ A and B

Q44 Which of the following is not a unit of viscosity?

   A   ❏ $kg\,m^{-1}\,s^{-1}$
   B   ❏ $g\,cm^{-1}\,s^{-1}$
   C   ❏ poise
   D   ❏ $kg\,cm^{-2}\,s^{-1}$
   E   ❏ none of the above

Q45 Which of the following methods may be used to measure viscosity?

   A   ❏ flow rheometry
   B   ❏ spectroscopy
   C   ❏ turbidity measurement
   D   ❏ particle size analysis
   E   ❏ scanning electron microscopy

Q46 Dilatant systems always exhibit the following properties:

   A   ❏ thixotropy
   B   ❏ increased viscosity as a function of increasing shearing stress
   C   ❏ decreased viscosity as a function of increasing shearing stress
   D   ❏ increased viscosity as a function of decreasing shearing stress
   E   ❏ constant viscosity at high shearing stresses

Q47   Creams may not exhibit the following rheological properties:

A   ❑ thixotropy
B   ❑ dilatancy
C   ❑ pseudo-plastic flow
D   ❑ decreased viscosity as a function of increasing temperature
E   ❑ none of the above

Q48   The following observations are consistent with pharmaceutical systems that exhibit plastic flow:

A   ❑ a yield stress is present
B   ❑ at stresses below the yield stress, flow does not occur
C   ❑ at stresses greater than the yield stress, Newtonian flow occurs
D   ❑ A and B
E   ❑ A, B and C

Q49   Highly concentrated suspensions frequently exhibit the following rheological properties:

A   ❑ Newtonian flow
B   ❑ plastic flow
C   ❑ pseudo-plastic flow
D   ❑ dilatant flow
E   ❑ B and C

Q50   Creep analysis involves the following:

1   ❑ exposure of the sample to a fixed stress, removal of the stress and subsequent monitoring of product recovery
2   ❑ increasing the rate of shear of the sample and monitoring changes in viscosity
3   ❑ monitoring the modulus of the pharmaceutical systems as a function of oscillatory stresses of increasing frequency
    A ❑ 1
    B ❑ 2
    C ❑ 3
    D ❑ 1 and 2
    E ❑ 1, 2 and 3

Q51   The following statements are true concerning non-destructive oscillatory tests:

A   ❑ these tests involve the application of small stresses or strains to a sample

B ❏ oscillatory rheology involves the application of a sinusoidal stress or strain to a sample

C ❏ oscillatory rheology determines the loss (G″) and storage (G′) moduli of a sample

D ❏ oscillatory tests directly measure the specific viscosity

E ❏ A, B and C

Q52 Product rheology is important in the following processes:

A ❏ drug release from an effervescent tablet

B ❏ spreading a topical product on to the skin

C ❏ mixing a cream during manufacture

D ❏ drug release from a gel

E ❏ B, C and D

Q53 Which of the following statements is incorrect regarding the rheological properties of polymer solutions?

A ❏ the viscosity of a polymer solution is dependent on the chemical structure of the polymer

B ❏ the viscosity of a polymer solution is dependent on the choice of solvent

C ❏ the viscosity of a polymer solution is dependent on the molecular weight of the polymer

D ❏ in general, the viscosity of solutions of hydrophobic polymers in hydrophobic solvents is greater than the viscosity of solutions of hydrophilic polymers in hydrophilic solvents

E ❏ in general, the viscosity of solutions of hydrophilic polymers in hydrophilic solvents is greater than the viscosity of solutions of hydrophobic polymers in hydrophobic solvents

Q54 Which of the following statements is true concerning the viscosity of polymer solutions?

A ❏ the relative viscosity is the ratio of the viscosity of the solvent to the ratio of the polymer solution

B ❏ the specific viscosity (η_{sp}) may be defined as $\eta_{relative}^{-1}$

C ❏ the reduced viscosity (η_{sp}/C) is independent of polymer concentration (C) in ideal solutions

D ❏ the intrinsic viscosity [η] may be defined as the reduced viscosity at zero concentration

E ❏ B, C and D

Q55   The viscosity of aqueous pharmaceutical gels is usually dependent on:

A   ❏ temperature
B   ❏ nature of the solvent
C   ❏ the presence of dissolved excipients, e.g. colours, preservatives
D   ❏ the molecular weight of the polymer used to form the gel
E   ❏ A, B and D

Q56   Surface tension of a liquid is measured using a:

A   ❏ Coulter counter
B   ❏ stalagmometer
C   ❏ pycnometer
D   ❏ Anderson's pipette
E   ❏ stage micrometer

Q57   The wetting ability of a vehicle can be detected by observing the:

A   ❏ contact angle
B   ❏ critical angle
C   ❏ angle of repose
D   ❏ interfacial angle
E   ❏ critical micellar concentration

Q58   Draves' test is associated with increasing the efficiency of:

A   ❏ detergents
B   ❏ wetting agents
C   ❏ suspending agents
D   ❏ antifoaming agents
E   ❏ preservatives

Q59   Chemically, spans are:

A   ❏ polyethylene glycol esters of fatty acids
B   ❏ polyoxyethylene sorbitan esters of fatty acids
C   ❏ sorbitan esters of fatty acids
D   ❏ polyethylene glycol esters of alcohols
E   ❏ polyoxyethylene sorbitan esters of alcohols

Q60   The potential between the surface of the tightly bound layer and the electro-neutral region is termed the:

A   ❏ Nernst potential
B   ❏ Stern potential
C   ❏ surface potential

D ❏ zeta potential

E ❏ streaming potential

Q61 At a certain temperature, aqueous solutions of many non-ionic surfactants become turbid. However, on cooling, clarity of the solution is restored. The characteristic temperature at which turbidity appears is known as the:

A ❏ cloud point

B ❏ Krafft temperature

C ❏ triple point

D ❏ critical micellar concentration

E ❏ dew point

Q62 Surface active agents used as preservatives include:

1 ❏ benzalkonium chloride

2 ❏ cetrimide

3 ❏ Tween 20

    A ❏ 1, 2 and 3

    B ❏ 1 and 2

    C ❏ 1 and 3

    D ❏ 2 and 3

    E ❏ 3

Q63 'Adsorption' phenomena are based on the:

1 ❏ Freundlich equation

2 ❏ Kozeny–Carman equation

3 ❏ Henderson–Hasselbalch equation

4 ❏ Langmuir equation

    A ❏ 1

    B ❏ 2

    C ❏ 1 and 2

    D ❏ 1 and 4

    E ❏ 3 and 4

Q64 An emulsion is formulated using the dry gum technique. The emulsion is stabilised using a blend of two surfactants, Tween 80 and Span 80, with hydrophilic lipophilic balance (HLB) values of 15 and 5, respectively. What would be the HLB value of a blend of equal amounts of two surfactants?

A ❏ 3

B ❏ 5

C ❑ 10
D ❑ 15
E ❑ 20

Q65 Tween 80 is used as:

A ❑ an oil-in-water emulsifier
B ❑ a water-in-oil emulsifier
C ❑ a viscosity modifier
D ❑ a deflocculating agent
E ❑ a flocculating agent

Q66 Which of the following is the correct order of the rate of hydrolysis?

A ❑ thioester < ester < amide
B ❑ thioester > ester > amide
C ❑ ester > thioester > amide
D ❑ amide > thioester > ester
E ❑ ester > amide > thioester

Q67 For chemical reactions between ions of opposite charge, the degradation rate:
❑ increases with an increase in the ionic strength of the buffer
❑ decreases with an increase in the ionic strength of the buffer
❑ increases with a decrease in the dielectric constant of the solvent
❑ increases with an increase in the dielectric constant of the solvent

A ❑ 1 and 3
B ❑ 2 and 3
C ❑ 1 and 4
D ❑ 2 and 4
E ❑ 1

Q68 Which of the following is not a synergist to antioxidants?

A ❑ citric acid
B ❑ tartaric acid
C ❑ phosphoric acid
D ❑ lecithin
E ❑ butylated hydroxytoluene

Q69 An oxidation reaction involves a:

1 ❑ gain of oxygen
2 ❑ loss of hydrogen
3 ❑ gain of electrons

4 ❏ loss of electrons
  A ❏ 1
  B ❏ 1 and 2
  C ❏ 1 and 3
  D ❏ 1, 2 and 3
  E ❏ 1, 2 and 4

Q70 The unit for the degradation rate constant of a zero-order reaction is:

A ❏ per unit time
B ❏ per unit concentration
C ❏ concentration per unit time
D ❏ per unit concentration per unit time
E ❏ concentration × time

Q71 The unit for the degradation rate constant of a first-order reaction is:

A ❏ per unit time
B ❏ per unit concentration
C ❏ concentration per unit time
D ❏ per unit concentration per unit time
E ❏ concentration × time

Q72 The unit for the degradation rate constant of a second-order reaction is:

A ❏ per unit time
B ❏ per unit concentration
C ❏ concentration per unit time
D ❏ per unit concentration per unit time
E ❏ concentration × time

Q73 A drug solution (100 mg/mL) degrades to 90 mg/mL in 1 day and to 50 mg/mL in 5 days. What is the order of the degradation reaction?

A ❏ zero order
B ❏ first order
C ❏ second order
D ❏ third order
E ❏ can not be determined with the information provided

Q74 A drug solution (100 mg/mL) degrades to 90 mg/mL in 1 day and to 50 mg/mL in 5 days. What is the likely value of the degradation reaction constant?

A ❏ 5
B ❏ 10

C ❏ 20
D ❏ 30
E ❏ 60

Q75 Which of the following chemicals is not an antioxidant used in pharmaceutical formulations?

A ❏ α-tocopherol
B ❏ propyl gallate
C ❏ butylated hydroxyanisole
D ❏ butylated hydroxytoluene
E ❏ none of the above

Q76 A drug solution (100 mg/mL) undergoes a first-order degradation reaction. If the half-life of the degradation reaction is equal to 30 days, what is the likely value of the degradation rate constant of the reaction?

A ❏ 3.33
B ❏ 1.66
C ❏ 0.83
D ❏ 0.41
E ❏ none of the above

Q77 A drug solution (100 mg/mL) undergoes a first-order degradation reaction. If the half-life of the degradation reaction is equal to 30 days, what is the shelf life of the solution (i.e. the time required to degrade to 90% of its initial concentration)?

A ❏ 10 days
B ❏ 6 days
C ❏ 4.5 days
D ❏ 3 days
E ❏ 1.5 days

Q78 A drug solution (100 mg/mL) undergoes a first-order degradation reaction. If the half-life of the degradation reaction is equal to 30 days, what is the number of days required to degrade to 25% of its initial concentration?

A ❏ 45
B ❏ 60
C ❏ 70
D ❏ 90
E ❏ 120

Q79 Which of the following are potential adverse effects of instability in pharmaceutical products?

1 ❑ loss of active ingredient
2 ❑ alteration of bioavailability
3 ❑ decline of microbial status
 A ❑ 1
 B ❑ 1 and 2
 C ❑ 1 and 3
 D ❑ 2 and 3
 E ❑ 1, 2 and 3

Q80 Which of the following is the most important mechanism responsible for solid-state drug degradation?

 A ❑ pyrolysis
 B ❑ photolysis
 C ❑ oxidation
 D ❑ solvolysis
 E ❑ leaching from packaging

Q81 Which of the following reasons can lead to failure of new molecules in the drug-discovery programme?

1 ❑ poor biopharmaceutical properties such as aqueous solubility, stability, permeability and bioavailability
2 ❑ insufficient efficacy
3 ❑ toxicity
 A ❑ 1 and 2
 B ❑ 2 and 3
 C ❑ 1 and 3
 D ❑ 1
 E ❑ 1, 2 and 3

Q82 The statistical average cost (in $US millions) for discovering and developing a new drug is:

 A ❑ 20–50
 B ❑ 100–200
 C ❑ 300–500
 D ❑ 800–1000
 E ❑ 2000–3000

Q83 Preformulation studies:

1 ❑ help in early termination of compounds likely to fail in the drug-development process
2 ❑ help in scientifically developing a formulation for new molecules
3 ❑ are a regulatory requirement

A ❑ 1
B ❑ 1 and 2
C ❑ 1 and 3
D ❑ 2 and 3
E ❑ 1, 2 and 3

Q84 Which of the following is not a 'fundamental' property of a drug candidate?

A ❑ solubility
B ❑ melting point
C ❑ photostability
D ❑ flow behaviour
E ❑ partition coefficient

Q85 Which of the following is not a 'derived' property of a drug compound?

A ❑ solubility
B ❑ bulk density
C ❑ compression characteristics
D ❑ flow behaviour
E ❑ microscopy

Q86 Which of the following would not be tested in a typical preformulation programme?

A ❑ solubility
B ❑ pK_a
C ❑ toxicology
D ❑ melting point
E ❑ partition coefficient

Q87 Which of the following techniques is (are) commonly used to examine polymorphism?

1 ❑ viscometry
2 ❑ solubility

3 ☐ microscopy
A ☐ 1
B ☐ 1 and 2
C ☐ 1 and 3
D ☐ 2 and 3
E ☐ 1, 2 and 3

Q88 A high degree of hygroscopicity can cause:
☐ stability (physical and chemical)-related problems on storage
☐ physical changes such as cake formation and colour changes
☐ processing-related problems such as poor flow and sticking

A ☐ 1
B ☐ 1 and 2
C ☐ 1 and 3
D ☐ 2 and 3
E ☐ 1, 2 and 3

Q89 In the drug-discovery programme, the usual percentage of new compounds that are discontinued because of poor solubility is:

A ☐ 5
B ☐ 20
C ☐ 40
D ☐ 70
E ☐ 90

Q90 In the Noyes–Whitney equation, $dC/dt = DA(C_s - C)/h$, A refers to:

A ☐ the effective surface area of drug particles in contact with gastrointestinal fluids
B ☐ the surface area of suspended particles in a suspension
C ☐ the angle of repose
D ☐ the area of emulsion droplets
E ☐ the combined area of a powder mix

Questions 91–93 involve the following case:

In a new drug-discovery research group, a pharmaceutical scientist is provided with 50 mg each of four new compounds believed to have potential anticancer activity. A decision needs to be made to evaluate and select one of the four compounds for further development as an intravenous dosage form.

Q91 Following is (are) the recommended preformulation test(s) to make the decision:

A ❑ solubility in normal saline
B ❑ pH solubility profile
C ❑ solution stability
D ❑ solubility in buffer of pH 7.4
E ❑ all of the above

Q92 Since the quantity of compounds available is very small, the best method to determine solubility in saline and buffer is:

A ❑ centrifugation study
B ❑ equilibrium solubility study
C ❑ viscosity
D ❑ gravimetry
E ❑ derivatisation

Q93 The water solubility of four compounds was found to be lower than the dose required for intravenous formulation. You will suggest the following:

A ❑ discontinue the compounds from further development
B ❑ attempt to increase the solubility by use of surfactants, pH manipulation, complexation or other applicable methods
C ❑ combine with another new compound
D ❑ combine with an existing drug
E ❑ make a slow-release formulation, to be administered by the intramuscular route

Q94 From the existing drugs, the highest percentage of drugs are:

A ❑ weak acids
B ❑ weak bases
C ❑ strong acids
D ❑ non-ionisable
E ❑ strong bases

Q95 Which of the following class of drugs is likely to be well absorbed from the stomach?

A ❑ acidic
B ❑ weakly basic
C ❑ strongly basic
D ❑ acidic and basic
E ❑ neutral

Q96 The ionisation constant pK$_a$ can be determined by the following methods:

A ❏ ultraviolet (UV) spectroscopy
B ❏ in silico predictions such as PALLAS
C ❏ solubility
D ❏ kinetic pH profile
E ❏ all of the above

Q97 The partition coefficient (log P) of a given compound is a ratio of its solubility:

A ❏ in oil and water
B ❏ in water and oil
C ❏ in water and an oily formulation
D ❏ in the pure form and in emulsion form
E ❏ of its tablet and cream formulations

Q98 In the human gastrointestinal tract, the most adverse environment with reference to drug stability is found in the:

A ❏ stomach
B ❏ small intestine
C ❏ large intestine
D ❏ buccal area
E ❏ oesophagus

Q99 During formulation development, a drug excipient compatibility study can be carried out by:

A ❏ differential scanning calorimetry
B ❏ isothermal stress testing
C ❏ differential thermal analysis
D ❏ mini-formulation study
E ❏ all of the above

Q100 The incompatibility of dicalcium phosphate anhydrous and dicalcium phosphate dihydrate is because of:

A ❏ oxidation
B ❏ reduction
C ❏ hydrolysis
D ❏ photolysis
E ❏ isomerisation

Answers

A1   A

The main objectives of dosage form design are to ensure reproducible product quality, user acceptability, and uniformity in bioavailability, but not so much in drug response, which can be affected by many factors, including genetic makeup, age, body size and the use of other drugs.

A2   D

The base prednisolone has low aqueous solubility and suitable lipophilicity to be absorbed orally as a tablet. It is also insoluble enough to be formulated as a suspension for intramuscular injection to achieve the slow release of drug for a sustained effect.

A3   B

The sodium phosphate salt of prednisolone is water-soluble and most suitable to be formulated as an eye drop and enema to achieve a local effect.

A4   C

Enteric-coated tablets are designed to prevent the release of drug until it reaches the intestine. The onset of action takes up to several hours, depending on the nature of the enteric coating: the onset of action of a tablet takes from minutes to hours, while that of an intramuscular injection takes minutes and intravenous injection seconds, as the latter two do not involve an absorption process.

A5   C

Particle size reduction can only improve the bioavailability of poorly aqueous-soluble drugs showing a dissolution rate-limiting step in the absorption process. Particle micronisation can also increase air adsorption or static charge, leading to wetting or agglomeration problems.

A6   E

Many drugs can exist in more than one crystalline form, called polymorphs. Polymorphs may differ substantially in their physical properties, such as density, melting point, solubility and dissolution rate. Polymorphic transitions can occur during milling, granulation, drying and compacting operations (e.g. milling of digoxin and spironolactone can cause polymorphic transition).

A7   E

Surfactants can be used as emulsifying agents, solubilising agents and wetting agents (e.g. sodium stearate, polysorbates, sorbitan esters). Some cationic surfactants are used as antiseptics (e.g. benzalkonium chloride, chlorhexidine).

A8   E

The energy required to solvate a crystalline solid is much greater than that required for an amorphous solid. Therefore, the amorphous form of a drug is always more soluble than the corresponding crystalline form. Generally, the anhydrous forms dissolve more rapidly in water than the hydrous forms. In a hydrate, water could hydrogen-bond between the drug molecules and tie the lattice together; this would give a much stronger, more stable lattice and thus a slower dissolution rate. Addition of a conjugate base makes the water basic, there is more of the ionised form of the weak acid present and its solubility will be higher.

A9   D

The enteric-coated dosage form will protect erythromycin in the stomach acid and release the drug in the small intestine. The prodrug, erythromycin estolate, has a low dissolution rate and will be more stable in gastric fluid. Administering the drug with meals will prolong the retention of the drug in the stomach and thus will increase the extent of drug degradation by gastric acid.

A10   B

Lozenges are tablets that are intended to be dissolved slowly in mouth for local soothing or therapeutic effects.

A11   B

The free energy, or Gibbs' free energy (ΔG), decreases during a spontaneously occurring process until an equilibrium position is reached, when no more energy can be made available, i.e. $\Delta G = 0$ at equilibrium.

A12   E

According to Fick's law of diffusion, the rate of diffusion is proportional to the concentration difference between the two sides of the diffusion layer (ΔC). ΔC is the difference in concentration of solution at the solid surface (C_1) and the bulk of the solution (C_2). At equilibrium, the solution in contact with the solid (C_1) will be saturated, i.e. C_1 will become the saturation concentration, C_s.

A13 D

The rate of dissolution increases with increase in temperature, particle surface area and diffusion coefficient.

A14 C

The size of solid particles will affect only the dissolution rate, and not the saturation concentration, which is affected by the temperature, nature of the dissolution medium and the crystalline form of the sold.

A15 A

The solubility of an electrolyte will be higher in a medium with a high dielectric constant that can reduce the attractive forces between the oppositely charged ions produced by the dissociation of an electrolyte. If alcohol is added to an aqueous solution of a sparingly soluble electrolyte, the solubility of the latter is decreased because alcohol lowers the dielectric constant of the solvent and ionic dissociation of the electrolyte becomes more difficult.

A16 C

Mercuric iodide (HgI_2) is not very soluble in water but is more soluble in a solution of potassium iodide because of the formation of a water-soluble complex, $K_2(HgI_2)$.

A17 E

Solubility can be expressed as molarity (number of moles of solute in 1 L of solvent), molality (number of moles in 1 kg of solvent) and mole fraction (number of moles of solute per total number of moles of solute and solvent).

A18 E

Solubility of solids usually increases with increase in temperature, but some show the opposite trend, e.g. $(CH_3COO)_2Ca.2H_2O$. Solubility of some solids first increases with increase in temperature, but at a certain temperature decreases with increase in temperature; e.g. the dissolution of $Na_2SO_4.10H_2O$ is an endothermic process – its solubility first increases with rise in temperature until $32.5°C$ is reached; above this temperature, the solid is converted into anhydrous Na_2SO_4, and the dissolution of this compound is exothermic.

A19   D

The objectives of performing the dissolution test are to predict the absorption pattern in vivo if dissolution is a rate-determining step in the absorption process; and to serve as a quality-control procedure in the manufacturing process. However, it is not used to determine the solubility of the drug, as the dissolution medium will never be saturated.

A20   B

Surfactants can form micelles in aqueous solutions with the hydrophilic head of the surfactant in contact with the surrounding solution and the hydrophobic tail of the surfactant sequestered in the micelle centre. Poorly soluble drugs can be entrapped or complexed in the hydrophobic centre of the micelles, producing an apparent increase in solubility.

A21   E

Weak acids have a pK_a of about 4–5. At pH values above the pK_a, the acid will be primarily ionised. At about 3 units difference, the acid is primarily dissociated (ionised). Acids are more soluble whenever the $pH > pK_a$. The reverse is true for weak bases.

A22   B

Amphoteric drug molecules may act as both acids and bases. At high pH values these molecules act as acids, and at low pH values they act as bases. The partition coefficient of these molecules is maximal at the isoelectric point (i.e. the pH where the molecule exhibits an overall neutral charge).

A23   C

Proteins are amphoteric macromolecules. Increasing concentration of proteins increases the viscosity of the solution. The presence of electrolytes decreases the protein solubility.

A24   E

In concentrated drug solutions ion–ion interactions occur, thereby reducing the effective concentration (i.e. the activity) of the drug in solution. Such solutions are hypertonic.

A25   B

Polyprotic drugs have more than one site of ionisation, e.g. diacids, and therefore have more than one pK_a.

A26   E

The diffusion of drugs in solution follows the Stokes–Einstein equation, $D = kT/6\pi\eta R$, where T is the temperature, η is the viscosity and R is the hydrodynamic radius (related to molecular weight). For small organic molecules (< 700 g/mol), molecular weight does not affect diffusion.

A27   E

Non-ionic surfactants do not affect solution pH, will lower surface tension and may increase drug solubility in aqueous solution.

A28   C

Osmotic pressure is a colligative property and increases with the presence of drug in solution.

A29   E

Basic drugs will interact with anionic species in solution, which in turn may lower their solubility. Kaolin (as a suspended particle) may interact with basic drugs, thereby lowering solubility.

A30   E

The stability of suspensions is increased by slowing the rate of sedimentation of suspended particles. The rate of sedimentation is described as follows:

$$dx/dt = [2r^2(\rho_s - \rho_1)g]/9\eta$$

Increased stability may be gained by reducing particle size (r) and increasing the viscosity of the suspension (η).

A31   D

The stability of emulsions is partly controlled by the viscosity of the external phase. The effect of surfactants on the stability of emulsions is due to their ability to form an elastic film around the droplets of the internal phase and to lower the interfacial tension between the two phases.

A32 B

Oil-in-water emulsions are electrically conductive (due to the external phase consisting of water). This emulsion type may be injected intravenously if the particle size is $< 4\,\mu m$.

A33 D

Controlled flocculation facilitates interactions between particles at the secondary minimum, resulting in a large sediment volume and rate.

A34 C

The elastic film helps in stabilisation of an emulsion by keeping two immiscible phases separated from each other.

A35 E

The stabilisation effect is due to multilayer adsorption of the dispersed particles at the oil/water interface and formation of a protective film.

A36 B

The zeta potential is the main contributor to the primary maximum and is affected by the presence of electrolytes.

A37 A

As per Stoke's law, the rate of sedimentation is directly related to the particle size of the suspended particles.

A38 C

A flocculated suspension usually produces a bulky sediment, which is easy to redisperse.

A39 B

Emulsions are biphasic, thermodynamically unstable systems that can flocculate.

A40 B

Lyophilic colloids are hydrophilic polymers that are water-soluble, will increase the viscosity of the external phase, have large molecular weights, and are used to stabilise emulsions and suspensions.

A41 B

The viscosity of Newtonian fluids is temperature-dependent. Newtonian flow is exhibited only by simple molecules in solution.

A42 E

In pseudo-plastic flow the applied stress is not proportional to the rate of shear. The viscosity is calculated as the relationship between shearing stress and the rate of shear. It is temperature-dependent and decreases as a function of increasing applied stress. Pseudo-plastic flow is illustrated by polymeric systems.

A43 A

The Mark–Houwink equation is as follows:

$$[\eta] = KM\alpha$$

where M refers to the molecular weight and $[\eta]$ is the intrinsic viscosity. It is used to calculate the molecular weight or intrinsic viscosity of dilute polymeric solutions.

A44 D

Viscosity is the ratio of shearing stress ($N\,m^{-2}$) to the rate of shear (s^{-1}). A newton is defined dimensionally as $kg\,m\,s^{-2}$. The poise is another unit of measurement of viscosity (centimetre–gram–second (cgs) units).

A45 A

Flow rheometry is the traditional method used to measure viscosity.

A46 B

In dilatant flow, the applied stress is not proportional to the rate of shear. The viscosity is calculated as the relationship between shearing stress and the rate of shear. It is temperature-dependent and increases as a function of increasing applied stress. Thixotropy is time-dependent flow.

A47 B

The major instability in creams tends to be cracking of the emulsion and separation of the two phases.

A48 E

In plastic flow a yield stress is present. At stresses greater than the yield stress, Newtonian flow occurs.

A49 D

Formulations such as toothpastes indicate dilatant flow. Yield stress is the minimum stress required before flow will occur on application of stress.

A50 A

Creep analysis measures the viscoelastic properties of pharmaceutical formulations. It involves the exposure of the sample to a fixed stress, removal of the stress and subsequent monitoring of product recovery.

A51 E

Non-destructive oscillatory tests measure the viscoelastic properties of pharmaceutical formulations (G', G'', etc.). They involve the application of small sinusoidal stresses or strains to a sample.

A52 E

Product rheology affects situations in which the sample is required to flow or in situations involving diffusional processes, e.g. drug release.

A53 D

The rheological properties of a polymer solution typically depend on the type and molecular weight of polymer and the solvent in which the polymer is dissolved.

A54 E

This question concerns the different descriptions of viscosity. The definitions are relevant only to Newtonian systems.

A55 E

The viscosity of pharmaceutical gels is dependent on temperature, the nature of the solvent, the molecular weight and the type of polymer, but generally not on the presence of dissolved excipients.

A56   B

A stalagmometer or drop pipette is a device used for the determination of surface tension by measurement of the weight of a drop of liquid (drop weight method) or by determination of the number of drops (drop count method). It is based on the principle that if a liquid is allowed to fall slowly through a capillary, then the liquid first forms a drop at the tip of the tube, which gradually increases in size and finally detaches when the weight of the drop equals the total surface tension at the circumference of the tube.

A57   A

The wetting of a solid involves the displacement of air from the solid surface in order to bring the surface of the solid into contact with the liquid surface. The contact angle is a quantitative measure of the wetting of a solid by a liquid. It is the angle between the tangent to the periphery of the point of contact with the solid, and the surface of the solid.

A58   B

Draves' test is used to determine the relative efficiency of wetting agents. In this test a weighed skein of cotton yarn is allowed to sink through a wetting solution in a 500 mL graduated cylinder. The time required for sinking is noted and compared. The shorter the time taken, the higher is the wetting efficiency.

A59   C

Spans are a class of non-ionic surfactant with emulsifying, dispersing and wetting properties. Spans are oily liquids derived from sorbitan (a derivative of sorbitol) esterified with fatty acid. Therefore, chemically spans are sorbitan esters of fatty acids.

A60   D

The zeta potential, or electrokinetic potential, is defined as the difference in the potential between the surface of the tightly bound layer and the electro-neutral region of the solution. It can also be defined as the work required to bring a unit charge from infinity to the surface of the particles. It governs the degree of repulsion between the adjacent ions of like charges.

A61   A

The cloud point is the temperature above which an aqueous solution of a water-soluble surfactant is no longer completely soluble, precipitating as a second phase and giving the fluid a cloudy appearance. Storing formulations at temperatures higher than the cloud point may result in phase separation and instability.

A62   B

Benzalkonium chloride and cetrimide are examples of cationic surfactants containing a quaternary ammonium group. Cationic surfactants have an active positive charge. They are adsorbed on the cell surface by electrostatic interaction, and as a result the cell surface loses its integrity. Thus, because of their bactericidal action, benzalkonium chloride and cetrimide are used as preservatives.

A63   D

The relationship between the pressure of the gas and the amount adsorbed at constant temperature is called the adsorption isotherm. This relationship has been expressed by the Freundlich and Langmuir equations. The Freundlich equation relates the concentration of a solute on the surface of an adsorbent to the concentration of the solute in the liquid with which it is in contact. The Langmuir adsorption isotherm describes quantitatively the build-up of a layer of molecules on an adsorbent surface as a function of the concentration of the adsorbed material in the liquid in which it is in contact.

A64   C

The HLB of a mixture of surfactants can be determined by the following formula:

$$HLB_{mixture} = HLB_1 \times \% \text{ in the mixture} + HLB_2 \times \% \text{ in the mixture}$$

The question states that an equal amount of two surfactants are blended, so the formula becomes:

$$HLB_{mixture} = (HLB \text{ of Tween } 80 \times 0.5 + HLB \text{ of Span } 80 \times 0.5)$$

So the calculated $HLB_{mixture} = (15 + 5)/2 = 10$.

A65   A

Polyoxyethylene sorbitan monooleate (Tween 80) is soluble in water, since a number of ethylene oxides are adducted by the hydroxyl groups of the sorbitan esters. They are hence used as emulsifying agents for oil-in-water emulsions.

A66   B

Due to the difference in the electronegativity, thioester is more susceptible to nucleophilic acyl substitution than ester and amide.

A67 B

For degradation reactions between reacting species of opposite charge, increasing the ionic strength will decrease the reaction rate; while decreasing the dielectric constant of the solvent will increase the rate of the reaction according to the equation

$$\log(k_{obs}) = \log(k_0) + 1.018z_Az_B$$

where k_0 is the rate constant at zero ionic strength, and z_A and z_B are the charges on the reacting species A and B. For reacting species of opposite charge, z_A could be positive and z_B could be negative. Therefore, increasing the ionic strength will decrease the observed degradation rate (k_{obs}).
For the effect of the dielectric constant of the solvent:

$$\log(k_{obs}) = \log(k_\infty) - K_{Z_AZ_B}/\xi$$

where k_∞ is the rate constant at infinite dielectric constant. For reaction between ions of opposite charge, decreasing the dielectric constant ξ will increase the degradation rate.

A68 E

Citric acid, tartaric acid, phosphoric acid and lecithin are all synergists to antioxidants, although the mechanisms of action of some synergists are not fully understood.

A69 E

Oxidation involves a gain of oxygen, a loss of hydrogen or a loss of electrons.

A70 C

The unit for the degradation rate constant of a zero-order reaction is concentration per unit time according to the following equations for zero-order reactions:

$$-dC/dt = k$$
$$k = (C_0 - C)/t$$

A71 A

The unit for the degradation rate constant of a first-order reaction is per unit time according to the following equations for first-order reactions:

$$-dC/dt = kC$$
$$k = (1/t)\ln(C_0/C)$$

A72   D

The unit for the degradation rate constant of a second-order reaction is per unit concentration per unit time according to the following equations for second-order reactions:

$$-dC/dt = kC^2$$
$$k = (1/t)(1/C - 1/C_0)$$

A73   A

Refer to the equation in Q70. When the concentration decreases by 10 mg/mL (i.e. $C_0 - C$) in one day, if it follows a zero-order reaction, $k = 10$ mg/mL/day; when the concentration decreases by 50 mg/mL (i.e. $C_0 - C$) in 5 days, if it follows a zero-order reaction, $k = 10$ mg/mL/day. The k values are identical from both estimations with the zero-order reaction equation; therefore the reaction is likely to follow a zero-order reaction.

A74   B

Refer to Q70–Q73.

A75   E

α-Tocopherol, propyl gallate, butylated hydroxyanisole (BHA) and butylated hydroxytoluene (BHT) are all used as antioxidants in pharmaceutical formulations.

A76   E

For a first-order degradation reaction, the degradation rate constant (k) is equal to $0.693/t_{1/2}$. In this case, it will be equal to $0.693/30 = 0.0231$ day^{-1}.

A77   C

For a first-order degradation reaction, the shelf life is equal to $0.105/k$, i.e. $0.105/0.0231 = 4.5$ days.

A78   B

The half-life is equal to 30 days (i.e. the time required to reduce the concentration by 50%). It will take another one half-life to reduce the concentration from 50% to 25%. Therefore, the total time required to reduce the concentration to 25% will be equal to two half-lives, i.e. 60 days.
Alternatively, $0.25C_0 = C_0 e^{-0.0231t}$, and $t = 60$ days.

A79   E

The potential adverse effects of instability in pharmaceutical products include loss of active ingredient, alteration of bioavailability and decline of microbial status. Alteration of bioavailability could be due to change in the crystalline form or size of the drug particles. Extended storage could also lead to microbial contamination of the product, i.e. microbial instability.

A80   D

Solvolysis is the most important reaction responsible for drug degradation in the solid state. When the solvent is water, the reaction is hydrolysis. Very often, the source of water is moisture from the environment. Oxidation often occurs only in a solvent, although oxygen may be able to oxidise a drug in the absence of a solvent. Photolysis may take place in the absence of a solvent, but the ability of light to penetrate a solid matrix may be limited. Pyrolysis is normally not an important mechanism for solid-state drug degradation; it may be more prominent when the drug is exposed to very high temperatures during processing.

A81   E

The failure rate for new chemical entities is very high for a number of reasons, as listed in the three options. In addition, synthetic complexities and market reasons can also lead to discontinuation of interesting compounds.

A82   D

The average cost for bringing a new drug to the market over the past few years is estimated at around $US800 million.

A83   E

Preformulation research helps in achieving all three objectives during new drug-discovery programmes.

A84   D

Flow behaviour is a derived property which can be altered in a number of ways, such as size reduction and addition of lubricants.

A85   A

Solubility is a fundamental property of a drug candidate. It can not be altered easily.

A86 C

Toxicology is usually independent of the preformulation programme, although it forms a component of the matrix required for go/no-go decisions with new active molecules.

A87 D

Microscopy, hot stage microscopy, differential scanning calorimetry (DSC), infrared (IR) spectroscopy, dissolution and solubility, and X-ray diffraction (XRD) are commonly used to study polymorphism.

A88 E

Hygroscopic compounds tend to exhibit all of the listed problems if appropriate precautions are not taken.

A89 C

Around 40% of new compounds are discontinued because of poor solubility.

A90 A

The Noyes–Whitney equation helps in understanding the effects on drug absorption of particle size, solubility, saturation solubility and thickness of diffusion layer.

A91 E

Because the target formulation is an intravenous solution, all the listed parameters will have to be evaluated in order to explore the feasibility of the formulation.

A92 B

Of the options listed, equilibrium solubility measurement is the only one that can be used for accurate solubility estimation.

A93 B

Exploring some options to improve solubility can solve the problem and save an active compound from being dropped off. In the early stages of drug discovery, alternative routes such as intramuscular are not explored because of uncertainty regarding bioavailability.

A94   B

Around 75% of existing drugs are weak bases.

A95   A

Acidic drugs are generally well absorbed from the stomach because of their existence in an un-ionised form in the acidic environment.

A96   E

The ionisation constant, which refers to the pH range in which a substance is least ionised, can be determined by all of the listed methods.

A97   A

The partition coefficient is a ratio of solubility in oil and water. It is also referred to as the octanol–water coefficient, which refers to the solubility in octanol as a representative of oils and water.

A98   A

Because of the highly acidic pH, enzymes and a residence time of 2–3 hours, the stomach offers the most adverse environment for drug stability.

A99   E

A drug excipient compatibility study is a useful tool to estimate stability of a drug in the presence of excipients and saves considerable time if the formulation degrades during stability studies.

A100   C

Anhydrous compounds are unstable in the presence of dihydrate compounds because of hydrolysis.

Test 2

Pharmacokinetics and biopharmaceutics

Omathanu P Perumal, Alison Haywood,

Beverley D Glass and Paul Chi-Lui Ho

Gastrointestinal tract physiology, absorption, biopharmaceutics

Bioavailability, physicochemical and dosage form factors

Dosage regimens, pharmacokinetics

Modified-release peroral dosage forms

Controlled drug release

Introduction

Oral dosage forms are widely used due to the convenience of drug administration. There are several steps a dosage form/delivery system has to undergo before it produces a therapeutic response. This can be explained by the LADMER system, which includes *liberation* of a drug from the dosage form, *absorption* of the drug, *distribution* of the drug, *metabolism* of the drug, *excretion* of the drug and finally the *response*. Biopharmaceutics deals with the study of physicochemical and physiological factors that influence the liberation and absorption of drugs from different dosage forms. Pharmacokinetics deals with the absorption, distribution, metabolism and excretion of a drug; the study of drug response is known as pharmacodynamics. In simple terms, biopharmaceutics is what the pharmaceutical scientist does to the drug, pharmacokinetics is what the body does to the drug, and pharmacodynamics is what the drug does to the body. Optimisation of biopharmaceutics and pharmacokinetic properties plays a significant role in the development of new drugs. This can be exemplified by the fact that 40% of drug candidates do not make it to market because of poor biopharmaceutical and pharmacokinetic properties. Drug solubility and permeability are the two most important biopharmaceutical properties that influence drug absorption and oral bioavailability. This led to the biopharmaceutics classification system (BCS), which classifies drugs into four classes based on their aqueous solubility and permeability. These two properties are determined by the drug's physicochemical properties, such as its chemical structure, molecular weight, pK_a, partition coefficient, crystal structure and particle size, among others. Drugs with good aqueous solubility and membrane permeability generally show good oral absorption and bioavailability, provided the drug is stable in the gastrointestinal tract and does not undergo first-pass metabolism in the liver. Several technologies have emerged to address the poor solubility and permeability of drugs. The important pharmacokinetic parameters that influence the biological performance of dosage forms are volume of distribution, half-life, clearance and fraction absorbed. The volume of distribution and clearance influence the drug's half-life, which in turn governs the frequency of drug administration. Clearance and fraction absorbed influence the bioavailability, which in turn determines the dose of a drug. Depending on how a drug is distributed into the body, different mathematical models can be used to characterise the drug disposition and estimate the pharmacokinetic parameters. In oral modified-release systems when or where the drug is released in the gastrointestinal tract is modified. These systems can be broadly divided into delayed-release systems and extended-release systems. In the case of delayed-release systems, the drug release is delayed but not sustained. In the case of extended-release systems, the drug release is sustained or controlled with respect to time, thus reducing the frequency of administration.

Questions

Q1   Drug A is a weak acid (pK_a 6.4) and has a log P (partition coefficient) of 2.06. Drug B is also a weak acid (pK_a 6.4) and has a log P of 0.89. According to the pH-partition hypothesis, which of the following will be true with regard to the absorption of these two drugs from the stomach?

A   ❑ absorption of Drug A > Drug B
B   ❑ absorption of Drug A = Drug B
C   ❑ absorption of Drug B > Drug A
D   ❑ only Drug A will be absorbed from the stomach
E   ❑ neither drug will be absorbed from the stomach

Q2 Which of the following salts will be affected by the common-ion effect in the stomach?

A ❏ sulfate
B ❏ maleate
C ❏ mesylate
D ❏ hydrochloride
E ❏ tosylate

Q3 Which of the drug absorption processes requires energy for drug transport?

A ❏ facilitated carrier-mediated transport
B ❏ active carrier-mediated transport
C ❏ passive drug diffusion
D ❏ convective solvent flow
E ❏ ion-pair transport

Q4 Which of the following is the main driving force for passive drug diffusion?

A ❏ concentration gradient
B ❏ partition coefficient
C ❏ drug solubility
D ❏ surface area available for drug diffusion
E ❏ membrane thickness

Q5 After oral administration, most drugs are absorbed from the:

A ❏ stomach
B ❏ small intestine
C ❏ large intestine
D ❏ oesophagus
E ❏ oral cavity

Q6 Enterohepatic recycling is seen mainly with drugs that are:

A ❏ metabolised in the liver before reaching the systemic circulation
B ❏ metabolised in the liver after reaching the systemic circulation
C ❏ excreted in the bile
D ❏ excreted in the urine
E ❏ absorbed in the stomach

Q7 Which of the following is considered a generic substitution for para-cetamol tablet?

A ❏ different brand of paracetamol tablet
B ❏ paracetamol suspension
C ❏ aspirin tablet
D ❏ ibuprofen tablet
E ❏ paracetamol-with-codeine tablet

Q8 Which of the following is considered a pharmaceutical alternative for tetracycline hydrochloride capsule?

1 ❏ different brand of tetracycline hydrochloride capsule
2 ❏ tetracycline phosphate capsule
3 ❏ tetracycline hydrochloride oral suspension
4 ❏ doxycycline hydrochloride capsule
  A ❏ 1
  B ❏ 1, 2 and 3
  C ❏ 4
  D ❏ 2 and 4
  E ❏ 2 and 3

Q9 Human bioequivalence studies are not required for:

A ❏ drugs with good water solubility
B ❏ drugs with poor membrane permeability
C ❏ drugs acting locally in the gastrointestinal tract
D ❏ drugs formulated in immediate-release oral dosage forms
E ❏ drugs that are unstable in the gastrointestinal tract

Q10 Which of the following is (are) not true for oral bioequivalence studies?

1 ❏ area under the curve (AUC) of the test and reference products are compared
2 ❏ maximum concentration (C_{max}) of the test and reference products are compared
3 ❏ bioequivalence studies are usually done in patients
4 ❏ oral bioavailability is compared with the bioavailability from intravenous injection
  A ❏ 1
  B ❏ 2, 3 and 4
  C ❏ 3

D ❑ 3 and 4
E ❑ 1, 2, 3 and 4

Q11 The biopharmaceutics classification system (BCS) is based on a drug's:

1   ❑ aqueous solubility
2   ❑ stability in the gastrointestinal tract
3   ❑ absorption mechanism
4   ❑ membrane permeability
A ❑ 1 and 2
B ❑ 2 and 4
C ❑ 1 and 4
D ❑ 1, 2, 3 and 4
E ❑ 1, 3 and 4

Questions 12–15 involve the following case:

An investigational drug is being developed for oral administration. The drug is stable in gastrointestinal fluids and has good water solubility and membrane permeability. The drug is a weak acid and has a pK_a of 4.2.

Q12 Where do you expect the drug to be absorbed based on the pH-partition hypothesis?

A   ❑ in the stomach
B   ❑ in the small intestine
C   ❑ in the large intestine
D   ❑ throughout the gastrointestinal tract
E   ❑ not absorbed from any region of the gastrointestinal tract

Q13 An in vitro experiment using everted rat intestine was used to study the mechanism of absorption of this drug. What information can be obtained from this experiment?

A   ❑ passive transport
B   ❑ carrier-mediated transport
C   ❑ enterohepatic recycling
D   ❑ first-pass metabolism
E   ❑ ion-pair-mediated absorption

Q14 The absolute bioavailability of an orally administered solution of this drug is 20%. What could be the possible reasons for the low oral bioavailability of this drug?

1 ❏ first-pass metabolism
2 ❏ drug is susceptible to P-glycoprotein efflux
3 ❏ drug exists in different polymorphs
4 ❏ drug is highly ionised throughout the gastrointestinal fluid
A ❏ 1
B ❏ 3
C ❏ 1 and 2
D ❏ 1, 2, 3 and 4
E ❏ 1, 3 and 4

Q15 Studies in human volunteers found that the solution formulation and suspension formulation of this drug had the same oral bioavailability. What can you conclude from this finding?

A ❏ drug is absorbed through intestinal transporters
B ❏ drug dissolution is slow
C ❏ drug does not undergo first-pass metabolism
D ❏ drug dissolution is very rapid
E ❏ drug is passively absorbed

Questions 16–20 involve the following case:

The data in Table 2.1 were generated for an investigational drug from studies in human volunteers.

Q16 What is the oral bioavailability of this drug?

A ❏ 0.8
B ❏ 1

Table 2.1 Bioavailability data from different routes of administration for an investigational drug

Formulation	Dose (mg/kg)	AUC (µg h/mL)
Intravenous solution	2	29
Oral solution	10	145
Oral tablet	20	232
Oral capsule	10	116

C   ❑ 1.25
D   ❑ 0.3
E   ❑ 0.2

Q17   Which of the following is the correct option regarding the oral bio-availability of this drug?

A   ❑ the drug exhibits first-pass elimination
B   ❑ the drug is susceptible to P-glycoprotein efflux
C   ❑ drug absorption is not solubility limited
D   ❑ drug absorption is not permeability limited
E   ❑ drug is unstable in the gastrointestinal tract

Q18   What is the drug's bioavailability from the tablet formulation?

A   ❑ 0.8
B   ❑ 1
C   ❑ 1.25
D   ❑ 0.3
E   ❑ 0.2

Q19   What is the drug's bioavailability from the capsule formulation?

A   ❑ 0.8
B   ❑ 1
C   ❑ 1.25
D   ❑ 0.3
E   ❑ 0.2

Q20   What is the rate-limiting step in the bioavailability of this drug from both the solid dosage forms?

A   ❑ first-pass elimination
B   ❑ dissolution rate
C   ❑ disintegration rate
D   ❑ drug permeability
E   ❑ drug stability in the gastrointestinal tract

Q21   The presence of food in the gastrointestinal tract may decrease the absorption of some drugs because food:

A   ❑ may increase the amount of surfactants in gastric juice
B   ❑ may increase the degree of agitation experienced by the drug particles

C ❏ may decrease the thickness of the diffusion layer around each drug particle

D ❏ may reduce the rate of diffusion through the diffusion layer

E ❏ none of the above

Q22 The antifungal drug ketoconazole, a weak base, is poorly absorbed when it is administered 2 h after the administration of cimetidine, an H_2 blocker, because:

A ❏ ketoconazole forms an insoluble complex with cimetidine

B ❏ cimetidine delays the gastric emptying of ketoconazole

C ❏ cimetidine reduces the gastrointestinal motility

D ❏ A, B and C

E ❏ none of the above

Q23 According to the pH-partition hypothesis, only the non-ionised forms of drugs can pass through the lipid gastrointestinal/blood barrier. Despite their high degree of ionisation, many weak acids are still absorbed quite well from the small intestine because:

1 ❏ the lipid gastrointestinal barrier is leaky to the permeation of small drug molecules

2 ❏ a large mucosal surface area is available for absorption in the small intestine

3 ❏ the effective pH at the surface of the intestinal mucosa is lower than the bulk pH in the lumen of the small intestine

A ❏ 1

B ❏ 2

C ❏ 1 and 2

D ❏ 2 and 3

E ❏ 1, 2 and 3

Q24 According to the Henderson–Hasselbalch equation, in gastric fluid with a pH of 1.2 the percentage of a weak acid ($pK_a = 3$) available in the un-ionised form will be equal to:

A ❏ 93.5

B ❏ 95.6

C ❏ 98.4

D ❏ 99.9

E ❏ none of the above

Q25 According to the Henderson–Hasselbalch equation, in blood of pH 7.4 the percentage of a weak acid ($pK_a = 3$) available in the ionised form will be equal to:

A   ❏ 93.5
B   ❏ 95.6
C   ❏ 98.4
D   ❏ 99.9
E   ❏ none of the above

Q26 According to the Henderson–Hasselbalch equation, in gastric fluid of pH 1.2 the percentage of a weak base ($pK_a = 5$) available in the ionised form will be equal to:

A   ❏ 93.5
B   ❏ 95.6
C   ❏ 98.4
D   ❏ 99.9
E   ❏ none of the above

Q27 According to the Henderson–Hasselbalch equation, in blood of pH 7.4 the percentage of a weak base ($pK_a = 5$) available in ionised form will be equal to:

A   ❏ 93.5
B   ❏ 95.6
C   ❏ 98.4
D   ❏ 99.9
E   ❏ none of the above

Q28 Which of the following statements about surfactants is incorrect?

A   ❏ surfactant monomers can potentially disrupt the integrity and function of a membrane and hence enhance drug penetration and absorption
B   ❏ drug absorption can be inhibited by incorporating the drug into surfactant micelles
C   ❏ release of poorly soluble drugs from tablets and hard gelatin capsules may be increased by the inclusion of surfactants in their formulations
D   ❏ the wetting effect may aid the penetration of gastrointestinal fluids into the solid dosage form and reduce the tendency of poorly soluble drug particles to aggregate in the gastrointestinal fluids
E   ❏ none of the above

Q29   Which of the following statements about viscosity-enhancing agents is incorrect? Viscosity enhancing agents may affect drug absorption by:

A   ❑ a decrease in gastric residence time
B   ❑ a decrease in intestinal motility
C   ❑ a decrease in dissolution rate of the drug
D   ❑ a decrease in the rate of movement of drug molecules to the absorbing membrane
E   ❑ none of the above

Q30   Which of the following drug complex(es) reduces the absorption of drug?

1   ❑ tetracycline and calcium ion
2   ❑ phenytoin and calcium sulfate dihydrate
3   ❑ ergotamine tartrate and caffeine
   A ❑ 1
   B ❑ 2
   C ❑ 3
   D ❑ 1 and 2
   E ❑ 1, 2 and 3

Q31   The absorption of which of the following drugs cannot be improved by particle size reduction?

A   ❑ griseofulvin
B   ❑ spironolactone
C   ❑ digoxin
D   ❑ erythromycin
E   ❑ none of the above

Q32   Polymorphs may differ substantially in:

1   ❑ dissolution rate
2   ❑ melting point
3   ❑ density
   A ❑ 1
   B ❑ 1 and 2
   C ❑ 2 and 3
   D ❑ 1 and 3
   E ❑ 1, 2 and 3

Q33   Which of the following formulation factors can not influence the bioavailability of drugs from hard gelatin capsules?

A   ❑ particle size of the drug

B   ❑ nature of the lubricant
C   ❑ conditions of the filling process
D   ❑ properties of the capsule shell
E   ❑ none of the above

Q34   Which of the following statements about the film coating of a tablet core with a film of hydroxypropylmethylcellulose is (are) correct?

A   ❑ the film will protect the drug from the gastric acid
B   ❑ the film will delay the rate of drug release
C   ❑ the film will influence the rate of drug absorption
D   ❑ A, B and C
E   ❑ none of the above

Q35   The saturation solubility of the drug in solution in the diffusion layer surrounding the dissolving particle can be increased by:

1   ❑ increasing the temperature
2   ❑ increasing the agitation rate
3   ❑ decreasing the particle size
4   ❑ increasing the volume of the bulk fluid

    A   ❑ 1
    B   ❑ 1 and 2
    C   ❑ 2 and 3
    D   ❑ 3 and 4
    E   ❑ 1, 2, 3 and 4

Q36   Which of the following is not true for drug measurement in body fluids?

A   ❑ generally, plasma is used for measuring drug concentration
B   ❑ whole blood is used for drugs that are bound to blood cells
C   ❑ generally, only the unbound drug concentration is measured in plasma
D   ❑ generally, only the protein-bound drug concentration is measured in urine
E   ❑ generally, the total drug concentration (both unbound and bound) is measured in plasma

Q37   The half-life of ibuprofen is 2.2 h. The elimination rate constant is:

A   ❑ $3.17\,h^{-1}$
B   ❑ $0.32\,h^{-1}$
C   ❑ $3.17\,mg/h$

D   ❑ 0.32 h
E   ❑ 2.2 h^{-1}

Q38 Which of the following pharmacokinetic models show(s) a distinct distribution phase?

1   ❑ one-compartment model – intravascular
2   ❑ one-compartment model – extravascular
3   ❑ two-compartment model – intravascular
4   ❑ two-compartment model – extravascular
A ❑ 1 and 3
B ❑ 2 and 4
C ❑ 3
D ❑ 4
E ❑ 3 and 4

Q39 The distribution half-life of digoxin is 35 min and the elimination half-life is 2 days. After administering the drug, how long will it take for the drug to be completely distributed in the tissues?

A   ❑ 35 min
B   ❑ 2 days
C   ❑ 9.01 days
D   ❑ 2.6 h
E   ❑ immediately

Q40 Which one of the following is true for a two-compartment model drug?

A   ❑ absorption rate constant < elimination rate constant < distribution rate constant
B   ❑ absorption rate constant > distribution rate constant > elimination rate constant
C   ❑ absorption rate constant < distribution rate constant < elimination rate constant
D   ❑ distribution rate constant > absorption rate constant > elimination rate constant
E   ❑ elimination rate constant > absorption rate constant > distribution rate constant

Questions 41–47 involve the following case:

See the plasma drug concentration profile in Figure 2.1.

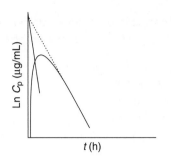

Figure 2.1   Plasma drug concentration profile

Q41   What does this plasma concentration–time profile represent?

A   ❏ intravascular drug administration – one-compartment model
B   ❏ extravascular drug administration – one-compartment model
C   ❏ intravascular drug administration – two-compartment model
D   ❏ extravascular drug administration – two-compartment model
E   ❏ intravascular drug administration – three-compartment model

Q42   Which of the following best describes the kinetics of the plasma concentration–time profile?

A   ❏ first-order absorption and zero-order elimination
B   ❏ zero-order absorption and first-order elimination
C   ❏ zero-order absorption and zero-order elimination
D   ❏ first-order absorption and first-order elimination
E   ❏ first-order absorption and zero-order distribution

Q43   Which of the following equations best describes the pharmacokinetics of this drug?

A   ❏ $C_p = C_0(e^{-kt} - e^{-k_a t})$
B   ❏ $C_p = C_0 e^{-kt}$
C   ❏ $C_p = Ae^{-\alpha t} + Be^{-\beta t} - C_0 e^{-k_a t}$
D   ❏ $C_p = Ae^{-\alpha t} + Be^{-\beta t}$
E   ❏ $C_p = [R/V_D k](1 - e^{-kt})$

Q44   What pharmacokinetic parameter can be obtained from the slope of the first straight line (solid line) in the graph?

A   ❏ elimination rate constant
B   ❏ absorption rate constant
C   ❏ drug clearance
D   ❏ volume of distribution
E   ❏ excretion rate constant

Q45 What pharmacokinetic parameter can be obtained from the slope of the second line (dotted line) in the graph?

A ❑ elimination rate constant
B ❑ absorption rate constant
C ❑ drug clearance
D ❑ volume of distribution
E ❑ initial drug concentration

Q46 What pharmacokinetic parameter can be obtained from the y-intercept?

A ❑ elimination rate constant
B ❑ absorption rate constant
C ❑ drug clearance
D ❑ volume of distribution
E ❑ initial drug concentration

Q47 Which of the following parameters can be used to determine the bioavailability of this drug?

A ❑ half-life
B ❑ absorption rate constant
C ❑ area under the curve (AUC)
D ❑ initial drug concentration
E ❑ elimination rate constant

Q48 Which of the following pharmacokinetic parameters is independent of the route of drug administration?

A ❑ area under the curve (AUC)
B ❑ volume of distribution (V_D)
C ❑ mean residence time
D ❑ half-life
E ❑ maximum concentration (C_{max})

Q49 The total clearance (CL_{tot}) is 11.76 L/h and the hepatic clearance (CL_h) is 10 L/h for aciclovir. What percentage of the drug will be excreted unchanged in the urine?

A ❑ 85
B ❑ 100
C ❑ 15
D ❑ 12
E ❑ 20

Q50 Which of the following is true for determining the volume of distribution (V_D) for a new drug?

A ❏ V_D is determined from single-dose oral drug administration

B ❏ V_D is determined from single-dose intravenous injection

C ❏ V_D is determined from single-dose subcutaneous drug administration

D ❏ V_D is determined from single-dose intramuscular drug administration

E ❏ V_D is determined from intravenous infusion

Q51 Which of the following statements is not true regarding plasma protein binding?

A ❏ drugs strongly bound to plasma proteins will generally have a small volume of distribution (V_D)

B ❏ drugs can alter the conformation of the protein binding site

C ❏ drugs can compete for the same binding site in a protein

D ❏ drugs can bind to more than one protein

E ❏ drugs strongly bound to plasma proteins will generally have a large V_D

Q52 The fraction unbound drug in plasma for a highly protein-bound drug is increased when:

1 ❏ there is competitive inhibition by another highly protein-bound drug

2 ❏ the plasma protein concentration increases

3 ❏ the plasma protein concentration decreases

4 ❏ there is allosteric inhibition by another drug

A ❏ 1
B ❏ 2
C ❏ 1 and 3
D ❏ 3 and 4
E ❏ 1, 3 and 4

Q53 A drug will have a long half-life if:

1 ❏ the volume of distribution (V_D) is large

2 ❏ drug clearance is high

3 ❏ it is highly bound to plasma proteins

A ❏ 1

B ❑ 2
C ❑ 1 and 2
D ❑ 1 and 3
E ❑ 2 and 3

Q54   Which of the following applies to flip-flop pharmacokinetics?

1   ❑ absorption rate constant $>$ elimination rate constant
2   ❑ absorption rate constant $<$ elimination rate constant
3   ❑ absorption rate constant $=$ elimination rate constant
4   ❑ absorption half-life $>$ elimination half-life
5   ❑ absorption half-life $<$ elimination half-life

A ❑ 1
B ❑ 2
C ❑ 3
D ❑ 2 and 4
E ❑ 1 and 5

Q55   Which of the following statements is true in non-linear pharma-cokinetics?

A   ❑ drug elimination follows first-order kinetics with increasing dose
B   ❑ drug elimination follows zero-order kinetics with increasing dose
C   ❑ half-life does not change with dose
D   ❑ clearance does not change with dose
E   ❑ area under the curve (AUC) is proportional to dose

Q56   The Michaelis–Menten constant V_{max} (the maximum rate of elimina-tion, i.e. elimination when all the enzyme-binding sites are saturated) for a drug is 15 mg/h. Which of the following dosing regimens will result in non-linear pharmacokinetics?

1   ❑ 15 mg/day
2   ❑ 300 mg twice daily
3   ❑ 50 mg four times daily
4   ❑ 150 mg three times daily

A ❑ 1
B ❑ 2, 3 and 4
C ❑ 4
D ❑ 2 and 4
E ❑ 1 and 3

Q57 The time to reach steady state during multiple dosing is determined by:

A ❏ area under the curve (AUC)
B ❏ volume of distribution (V_D)
C ❏ half-life
D ❏ maximum concentration (C_{max})
E ❏ time to maximum concentration (T_{max})

Q58 Which of the following parameters is useful for calculating the maintenance dose?

A ❏ clearance
B ❏ protein binding
C ❏ metabolic rate constant
D ❏ distribution half-life
E ❏ area under the curve (AUC)

Q59 Which of the following parameters is useful for calculating the loading dose?

A ❏ clearance
B ❏ volume of distribution (V_D)
C ❏ elimination rate constant
D ❏ elimination half-life
E ❏ area under the curve (AUC)

Q60 Ampicillin has a half-life of 0.9 h. If 250 mg of ampicillin capsules are administered for a week, how long will it take to reach steady state?

A ❏ 0.9 h
B ❏ 4 h
C ❏ 3 h
D ❏ 7 days
E ❏ immediately after administration

Q61 Which one of the following dosing frequencies (τ) will result in minimal drug accumulation in the body?

A ❏ $\tau = $ half-life ($t_{1/2}$)
B ❏ $\tau < 10\, t_{1/2}$
C ❏ $\tau > 10\, t_{1/2}$
D ❏ $\tau < 4.5\, t_{1/2}$
E ❏ $\tau > 4.5\, t_{1/2}$

Q62 Verapamil 180 mg tablets are prescribed three times daily to a cardiac patient for 1 month. If the patient misses a dose and 5 min remain until the next dose is due, what should the patient do?

A   ❑ take the missed dose
B   ❑ skip the missed dose
C   ❑ take two tablets for the next dose
D   ❑ skip all the remaining doses
E   ❑ take two tablets for the next dose

Q63 If the dosing frequency is equal to the half-life, which of the following statements is true?

A   ❑ the loading dose should be equal to the maintenance dose
B   ❑ the loading dose should be two times the maintenance dose
C   ❑ the loading dose should be three times the maintenance dose
D   ❑ the loading dose should be half of the maintenance dose
E   ❑ the loading dose should be three times lower than the maintenance dose

Questions 64–71 involve the following case:

A drug is given as multiple intravenous injections (200 mg every 4 h) for 4 days. The drug follows a one-compartment model. The clearance and volume of distribution (V_D) for this drug are 7.1 L/h and 44.9 L, respectively.

Q64 Calculate the elimination rate constant (h^{-1}) for this drug:

A   ❑ 0.08
B   ❑ 0.32
C   ❑ 0.16
D   ❑ 1.6
E   ❑ 0.8

Q65 Calculate the expected maximum concentration (C_{max}; mg/L) after the fourth dose:

A   ❑ 8.71
B   ❑ 4.60
C   ❑ 4.17
D   ❑ 5.87
E   ❑ 9.41

Q66 Calculate the expected minimum concentration (C_{min}; mg/L) after the fourth dose:

A ❑ 8.71
B ❑ 4.60
C ❑ 4.17
D ❑ 6.17
E ❑ 9.41

Q67 Calculate the expected maximum concentration (C_{max}; mg/L) at steady state:

A ❑ 8.71
B ❑ 4.82
C ❑ 6.17
D ❑ 5.87
E ❑ 9.41

Q68 Calculate the expected minimum concentration (C_{min}; mg/L) at steady state:

A ❑ 8.71
B ❑ 4.82
C ❑ 6.17
D ❑ 4.96
E ❑ 9.41

Q69 Calculate the plasma concentration (mg/L) 2 h after the third dose:

A ❑ 8.71
B ❑ 4.60
C ❑ 4.97
D ❑ 5.87
E ❑ 9.41

Q70 If the second dose was missed, what will be the plasma concentration (mg/L) 2 h after the third dose?

A ❑ 8.71
B ❑ 4.82
C ❑ 4.17
D ❑ 6.17
E ❑ 9.41

Q71 If the second dose was given 1 h later, what will be the plasma concentration (mg/L) 2 h after the third dose?

A ❏ 8.71
B ❏ 4.82
C ❏ 4.17
D ❏ 6.17
E ❏ 9.41

Q72 Which of the following is applicable for oral extended-release dosage forms?

A ❏ the drug release rate is faster than the absorption rate
B ❏ the drug release rate is slower than the absorption rate
C ❏ the drug release rate is equal to the absorption rate
D ❏ the drug release rate is equal to the elimination rate
E ❏ the absorption rate is equal to the elimination rate

Q73 Which drug molecules are suitable for oral extended-release formulations?

1 ❏ drugs with a half-life of 2–6 h
2 ❏ drugs with poor aqueous solubility
3 ❏ drugs that are uniformly absorbed throughout the gastrointestinal tract
4 ❏ drugs that have a narrow therapeutic index

A ❏ 1
B ❏ 1 and 3
C ❏ 3
D ❏ 2, 3 and 4
E ❏ 1, 3 and 4

Q74 Which of the following statements is true for enteric-coated systems?

A ❏ enteric-coated systems sustain the drug release
B ❏ enteric-coated systems release the drug immediately after administration
C ❏ enteric-coated systems release the drug only in the stomach
D ❏ enteric-coated systems release the drug only in the intestine
E ❏ enteric-coated systems release the drug in both the stomach and the intestine

Q75 Which of the following oral modified-release systems show Higuchi release kinetics?

A ❑ water-soluble matrix systems
B ❑ water-insoluble matrix systems
C ❑ water-soluble membrane controlled-release systems
D ❑ water-insoluble membrane controlled-release systems
E ❑ ion-exchange resins

Q76 Which of the following is true about surface-erosion-based drug-delivery systems?

A ❑ the rate of water diffusion into the system is greater than the rate of dissolution of the polymer
B ❑ the rate of water diffusion into the system is slower than the rate of dissolution of the polymer
C ❑ the rate of water diffusion into the system is equal to the rate of dissolution of the polymer
D ❑ the size of the delivery system increases with time
E ❑ the size of the delivery system stays constant

Q77 Which of the following can achieve zero-order drug release?

1 ❑ enteric-coated tablets
2 ❑ repeat-action tablets
3 ❑ osmotic controlled-release systems
4 ❑ ion-exchange resins
  A ❑ 4
  B ❑ 1 and 3
  C ❑ 3
  D ❑ 2, 3 and 4
  E ❑ 1, 3 and 4

Q78 How does a cationic ion-exchange resin release the drug in the gastrointestinal tract?

1 ❑ by exchanging with hydrogen ions in the stomach
2 ❑ by exchanging with sodium ions in the intestine
3 ❑ by exchanging with chloride ions in the stomach
4 ❑ by exchanging with carbonate ions in the intestine
  A ❑ 1
  B ❑ 2
  C ❑ 1 and 2
  D ❑ 3
  E ❑ 3 and 4

Q79 The drug release from oral osmotic-controlled release systems is affected by:

A ❑ pH in the gastrointestinal tract
B ❑ enzymes in the gastrointestinal tract
C ❑ water in the gastrointestinal tract
D ❑ ions in the gastrointestinal tract
E ❑ intestinal microflora

Q80 Which of the following oral delivery systems has the least chance for dose dumping?

A ❑ water-insoluble matrix systems
B ❑ water-soluble matrix systems
C ❑ multiparticulate systems
D ❑ osmotic controlled-release systems
E ❑ ion-exchange resins

Q81 Which of the following statements is true for osmotic controlled-release systems?

A ❑ in elementary osmotic systems the drug and osmogen layers are separate
B ❑ in push–pull osmotic systems the drug and osmogen layers are separate
C ❑ in push–pull osmotic systems the drug and osmogen layers are together
D ❑ in controlled-porosity osmotic systems the drug and osmogen layers are separate
E ❑ liquid drugs can not be delivered from osmotic controlled-release systems

Questions 82–86 involve the following case:

The physicochemical and pharmacokinetic properties of five new anti-hypertensive drugs are given in Table 2.2. All of these drugs exhibit similar pharmacological activity and have similar side-effects.

Q82 Which of the drugs exhibit the desirable solubility for developing an oral sustained-/controlled-release system?

A ❑ Drug A
B ❑ Drug B
C ❑ Drug C

Table 2.2 Physicochemical and pharmacokinetic properties of five new antihypertensive drugs

Property	Drug A	Drug B	Drug C	Drug D	Drug E
Aqueous solubility (mg/L)	10	0.2	0.01	1000	0.5
Absorption rate constant (h^{-1})	0.3	0.5	0.01	1.5	0.1
Absorption mechanism	Passive transport	Active transport	Passive transport	Active transport	Passive transport
Oral bioavailability (%)	80	75	20	85	55
Biological half-life (h)	4	3	1	20	6
Dose (mg)	100	20	1000	50	800

D   ❑ Drug D
E   ❑ Drug E

Q83 Which of the drugs exhibit the desirable absorption characteristics for developing an oral sustained-/controlled-release system?

A   ❑ Drug A
B   ❑ Drug B
C   ❑ Drug C
D   ❑ Drug D
E   ❑ Drug E

Q84 Which of the drugs may not warrant a sustained-/controlled-release system?

A   ❑ Drug A
B   ❑ Drug B
C   ❑ Drug C
D   ❑ Drug D
E   ❑ Drug E

Q85 With respect to the dose, which of the drugs is least preferred for an oral sustained-/controlled-release system?

A   ❑ Drug A
B   ❑ Drug B
C   ❑ Drug C
D   ❑ Drug D
E   ❑ Drug E

Q86 Which of the drugs has the optimal characteristics for developing a sustained-/controlled-release system?

A ❑ Drug A
B ❑ Drug B
C ❑ Drug C
D ❑ Drug D
E ❑ Drug E

Q87 Which of the following statements about the kinetics of drug release from a 'real' modified-release dosage form is incorrect?

A ❑ the initial dose is released rapidly
B ❑ the maintenance dose is released slowly to achieve a therapeutic concentration that is prolonged
C ❑ modified-release formulations use a chemical or physical barrier to provide rapid release
D ❑ coatings, wax, plastic matrices, microencapsulation, binding to ion-exchange resins, and osmotic pumps are examples of these chemical and physical barriers
E ❑ the priming dose is achieved by uncoated, rapidly releasing granules or pellets

Q88 Which of the following is a disadvantage of modified-release dosage forms?

A ❑ maintenance of therapeutic plasma drug concentrations
B ❑ improved patient compliance
C ❑ reduction in the incidence of localised gastrointestinal side-effects
D ❑ they contain larger amounts of drug than a single dose and therefore have the potential for an unsafe overdose
E ❑ cost savings due to better disease state management

Q89 Which of the following statements about factors influencing modified-release dosage forms is incorrect?

A ❑ residence time of the dosage form in the gastrointestinal tract is not influenced by stomach emptying
B ❑ drugs should ideally have a biological half-life of 4–6 h so that accumulation does not occur
C ❑ drugs should not have any significant form of pharmacologically inactive metabolites by first-pass metabolism

D ❑ residence time of the dosage form in the gastrointestinal tract is influenced by intestinal transit time

E ❑ dosage should not exceed 325 mg

Q90 Physicochemical properties of the drug that do not affect the release of the drug from a modified-release delivery system are:

1 ❑ aqueous solubility/partition coefficient of the drug
2 ❑ therapeutic index of the drug
3 ❑ salt form of the drug
A ❑ 1, 2 and 3
B ❑ 1 and 3
C ❑ 2 and 3
D ❑ 3
E ❑ 2

Q91 The two basic mechanisms controlling the release of a drug from a modified-release dosage form are dissolution of the drug and diffusion of the dissolved species. As these mechanisms apply to modified-release dosage forms, the following processes operate:

1 ❑ dehydrating the device and dissolution of water into the device
2 ❑ diffusion of the drug
3 ❑ diffusion of the dissolved drug out of the device
A ❑ 1, 2 and 3
B ❑ 1 and 2
C ❑ 2 and 3
D ❑ 1
E ❑ 3

Q92 Formulation components of a modified delivery system that does not directly control the release of the active drug from the system include:

A ❑ matrix formers
B ❑ channelling agents
C ❑ membrane formers
D ❑ solubilisers
E ❑ lubricants

Q93 Modified-release drug-delivery systems may be classified as inert, lipid or hydrophilic, depending on the nature of the excipients used. Which of the following is an example of a hydrophilic excipient?

A ❏ methacrylate copolymer
B ❏ carbopol
C ❏ polyvinyl acetate
D ❏ carnauba wax
E ❏ hydrogenated vegetable oils

Q94 Monolithic matrix systems include those in which the drug particles are dissolved in an insoluble matrix. Lipid matrices (not in common use now) and insoluble polymer matrices are examples of these systems. Which of the following statements about these matrices is incorrect?

A ❏ the drug is embedded in an inert polymer that is soluble in the gastrointestinal tract
B ❏ the release rate of the drug depends on the drug diffusing through a network of capillaries
C ❏ there is concern about initiators and catalysts used in the preparation of the polymer matrix leaching out with the drug
D ❏ since the matrices remain intact during gastrointestinal tract transit, this has led to concerns about impaction and patient concern when seeing the matrix as a 'ghost' in the stool
E ❏ the release rate of the drug may be modified by changing the porosity of the matrix

Q95 The role of a channelling agent in an insoluble polymer matrix is to:

A ❏ control the drug release from the matrix and as such is hydrophobic, solid at room temperature and does not melt at body temperature
B ❏ enhance the dissolution of the drug, with examples including polyethylene glycols, polyols and surfactants
C ❏ improve the flow properties of formulations during tabletting
D ❏ leach out from the formulations and be insoluble in the gastrointestinal tract
E ❏ create capillaries through which the dissolved drug may diffuse to be released and be water-soluble, with examples including sodium chloride and sugars

Q96 Hydrophilic colloid matrix systems are also called swellable soluble matrices. The principle of design of these hydrophilic matrices does not involve:

A ❏ a system comprising a mixture of drug, hydrophilic colloid, any release modifiers and a lubricant/glidant

B ❏ swelling of the hydrophilic colloid components on contact with water to form a hydrated matrix layer

C ❏ eroding the outer matrix such that it becomes more concentrated

D ❏ the nature of the colloid, which determines the rate of erosion

E ❏ diffusion of drug through the hydrated matrix, which controls its rate of release

Q97 There are two types of hydrophilic matrix, true gels and viscous or viscolised matrices. When considering these two systems, the following statement is true:

A ❏ the diffusion pathway is via the continuous phase in the interstices for viscous matrices

B ❏ there are no 'fixed' cross-links in viscous matrices

C ❏ the bulk viscosity of the true gel is not derived from the cross-linked polymer chains

D ❏ bulk viscosity correlates well with diffusion for true gels

E ❏ for true gels, diffusion does not correlate with microviscosity

Q98 The advantage of a hydrophilic matrix system includes:

A ❏ release of the drug is dependent on two diffusion processes

B ❏ erosion of the outer layer can complicate the release of the drug

C ❏ the system is capable of sustaining high drug loadings

D ❏ requirement of batch-to-batch consistency in the matrix-forming materials, other components and process parameters is difficult to achieve

E ❏ need for optimal rate-controlling polymers for different active drugs

Q99 Gel modifiers are materials that are incorporated into a hydrophilic matrix to modify the diffusional characteristics of the gel layer, to enhance drug diffusion and release of the drug. Which of the following statements about the functions of these gel modifiers is incorrect?

A ❏ allow more complete, more uniform hydration of the gel matrix

B ❏ allow more rapid hydration of the gel matrix

C ❏ dissociate from the matrix molecules and thus do not influence interactions at the molecular level, e.g. cross-linking

D ❏ modify the environment in the interstices of the gel, either to speed up or slow down diffusion

E ❏ suppress or promote the ionisation of ionisable polymers

Q100 The event that does not follow the immersion of a hydrophilic matrix containing a water-soluble drug in an aqueous medium is:

A ❏ surface drug does not dissolve and there is no 'burst' effect

B ❏ hydrophilic polymer hydrates and an outer gel layer forms

C ❏ gel layer becomes a barrier to the uptake of further water and the transfer of drug

D ❏ drug release occurs by diffusion through the gel layer

E ❏ following erosion, the new surface layer becomes hydrated and forms a new gel layer

Answers

A1   A

Both the drugs will be un-ionised at the acidic pH in the stomach. Drug A has a higher log P than Drug B. Therefore, Drug A will show a higher absorption than Drug B.

A2   D

A hydrochloride salt will ionise in solution, as shown in the equation below, but in gastric fluid the presence of chloride ions suppresses the drug ionisation, as shown by the thicker arrow in the reverse direction in the second equation, to maintain an equilibrium between the ionised and un-ionised form of the drug. This results in reduced solubility of the pharmaceutical salt.

$$DH^+Cl^- \rightleftharpoons DH^+ + Cl^-$$
$$DH^+Cl^- \rightleftharpoons DH^+ + Cl^-$$

A3   B

Active carrier-mediated transport requires energy for transport. The energy is provided by the breakdown of adenosine triphosphate (ATP) in the intestinal cells.

A4   A

According to Fick's law of diffusion, the concentration difference between the luminal fluid and the blood provides the main driving force for passive drug diffusion across the gastrointestinal membrane.

A5   B

The small intestine provides a large surface area for drug absorption due to the presence of villi. Therefore, most of the drugs are absorbed from the small intestine.

A6   C

Some drug (e.g. morphine) metabolites such as glucuronide conjugates are excreted in bile. The bile empties into the intestine and the drug is eventually excreted in faeces. The glucuronide conjugates can be converted back into the parent drug molecule by glucuronidase. This enzyme is produced by the intestinal microflora. If the regenerated drug is lipophilic, then it can be absorbed again through the intestinal membrane; this process is known as enterohepatic recycling.

A7   A

Generic substitution is dispensing a pharmaceutical equivalent, i.e. the same drug in a similar dosage form from a different manufacturer.

A8   E

Pharmaceutical alternative is substituting one dosage form with a different dosage form of the same drug or a different salt form of the same drug.

A9   C

A human bioequivalence study compares the biological performance of the drug products by measuring the drug concentration in plasma or urine. This requirement is waived for drugs that act locally in the gastrointestinal tract. In these cases the products can be compared using appropriate in vitro studies. For example, acid-neutralising capacity is a suitable in vitro test for comparing performance of antacid products.

A10   D

Bioequivalence studies are usually done in healthy volunteers. Oral bio-equivalence studies compare the bioavailability of test oral product with the oral bioavailability of a reference oral product (usually the existing product in the market).

A11   C

The BCS system classifies drugs into four classes based on the drug's solubility and permeability. Class I drugs are highly soluble and highly permeable through the gastrointestinal membrane. Class II drugs are poorly soluble but highly permeable through the gastrointestinal membrane. Class III drugs are highly soluble but poorly permeable through the gastrointestinal membrane. Class IV drugs are poorly soluble and poorly permeable through the gastro-intestinal membrane.

A12   A

Since the drug is a weak acid with a pK_a of 4.2, it will be predominantly un-ionised in the stomach pH of 1–2. According to the pH-partition hypothesis, the degree of ionisation will influence the drug absorption from different regions of the gastrointestinal tract. Generally, the un-ionised form of a drug has a higher membrane permeability than the ionised form of the drug.

A13   B

In everted rat intestine studies, the isolated rat intestine is used. The basolateral side (membrane that faces the blood) is kept in contact with the drug while the apical side (luminal side of the membrane) is kept in contact with the buffer. In a separate experiment, the drug is kept in contact with the apical side of the isolated intestine and the basolateral side facing the buffer. The transporters that are involved in carrier-mediated absorption are usually expressed on the apical side of the membrane. Therefore, the drugs that are absorbed by the transporters will result in lower permeation in the everted rat intestine experiment compared with the normal rat intestine experiment.

A14   C

The drug has good water solubility and membrane permeability and therefore the drug absorption is neither solubility- nor permeability-limited. However, the drug may be susceptible to first-pass metabolism in the liver before reaching the systemic circulation. Drugs absorbed from the intestine are taken up by the portal vein and emptied into the liver before reaching the systemic circulation. The drug can be converted to metabolites, thus reducing the oral bioavailability of the drug. Alternatively, during absorption through the intestinal membrane, the drug may be susceptible to P-glycoprotein efflux transporter expressed by the intestinal epithelial cells. This can transport the drug back into the intestine, resulting in reduced oral bioavailability.

A15   D

Unlike a solution dosage form, the drug has to dissolve and go into solution before it can be absorbed from the suspension dosage form. Therefore, the only difference between the two dosage forms is the drug dissolution rate from the suspension dosage form. Since the bioavailability of both these dosage forms is the same, the dissolution of the drug from the suspension appears to be very rapid.

A16   B

For determining the oral bioavailability of a drug, generally oral solution is compared with intravenous injection. The formula for absolute bioavailability is

$$\{[AUC]_{oral}/(Dose_{oral})\}/\{[AUC]_{IV}/(Dose_{IV})\} = (145/10)/(29/2) = 1$$

A17   D

The 100% bioavailability of the oral solution indicates that the drug has very good permeability through the gastrointestinal membrane. Since the drug is in solution form, the influence of drug solubility on drug absorption can not be commented upon.

A18   A

For determining the relative bioavailability, the test formulation is compared with the oral solution. The formula for relative bioavailability is

$$\{[AUC]_{tablet}/(Dose_{tablet})\}/\{[AUC]_{solution}/(Dose_{solution})\} = (232/20)/(145/10) = 0.8$$

A19   A

For determining the relative bioavailability, the test formulation is compared with the oral solution. The formula for relative bioavailability is

$$\{[AUC]_{capsule}/(Dose_{capsule})\}/\{[AUC]_{solution}/(Dose_{solution})\} = (116/10)/(145/10) = 0.8$$

A20   B

Compared with the oral solution, the solid dosage forms have to undergo dissolution before absorption. Unlike the capsule, the tablet has to undergo disintegration before dissolution. Since the bioavailability of the capsule and tablet dosage forms is the same, the drug dissolution is the only rate-limiting factor that can affect the bioavailability of these two solid dosage forms.

A21   D

The presence of food in the gastrointestinal tract increases the viscosity of the fluids. Hence, it may cause a decrease in the dissolution of a drug by reducing the rate of diffusion through the diffusion layer surrounding each dissolving drug particle and may decrease the absorption of some drugs. All the other factors could improve drug absorption by increasing drug dissolution.

A22   E

Ketoconazole is a weak base and is particularly sensitive to gastric pH. Cimetidine reduces gastric acid secretion, thus increasing the gastric pH, leading to the formation of the poorly soluble un-ionised ketoconazole with a significantly reduced bioavailability.

A23   D

Despite their high degree of ionisation, many weak acids are still absorbed fairly well from the small intestine because a large mucosal surface area is available for absorption in the small intestine and the effective pH at the surface of the intestinal mucosa is lower than the bulk pH in the lumen of the small intestine.

A24   C

According to the Henderson–Hasselbalch equation for a weak acid:

$pH = pK_a + \log([A^-]/[HA])$

$\log([A^-]/[HA]) = pH - pK_a = 1.2 - 3 = -1.8$

$[A^-]/[HA] = \text{antilog}(-1.8) = 0.016$

Therefore, the percentage of the acid available in the un-ionised form [HA] in the gastric fluid of pH 1.2 is equal to 98.4%.

A25   D

According to the Henderson–Hasselbalch equation for a weak acid:

$pH = pK_a + \log([A^-]/[HA])$

$\log([A^-]/[HA]) = pH - pK_a = pH - pK_a = 7.4 - 3 = 4.4$

$[A^-]/[HA] = \text{antilog}(4.4) = 25119$

Therefore, the percentage of the acid available in the ionised form [A$^-$] in blood of pH 7.4 is equal to 99.9%.

A26   D

According to the Henderson–Hasselbalch equation for a weak base:

$pK_a = pH + \log[BH^+]/[B]$

$\log[BH^+]/[B] = pK_a - pH = 5 - 1.2 = 3.8$

$[BH^+]/[B] = 6309.6$

Therefore, the percentage of the base available in the ionised form [BH$^+$] will be equal to 99.98%.

A27   E

According to the Henderson-Hasselbalch equation for a weak base:

$$pK_a = pH + \log[BH^+]/[B]$$

$$\log[BH^+]/[B] = pK_a - pH = 5 - 7.4 = -2.4$$

$$[BH^+]/[B] = 0.004$$

Therefore, the percentage of the base available in the ionised form $[BH^+]$ will be equal to 0.40%.

A28   E

Drug absorption can be inhibited by incorporating the drug into surfactant micelles. If such surfactant micelles are not absorbed, then the concentration of free drug available for absorption will be decreased. Inhibition of drug absorption is expected in the case of drugs that are normally soluble in the gastrointestinal fluids in the absence of surfactant.

A29   A

Viscosity-enhancing agents may affect drug absorption by decreasing gastric emptying time, thus causing an increase in gastric residence time.

A30   D

Both tetracycline-calcium ion and phenytoin-calcium sulfate dehydrate form insoluble complexes, thus reducing the drug absorption, but caffeine forms a more soluble complex with ergotamine tartrate, thus increasing the dissolution rate of the drug and enhancing the drug absorption.

A31   D

Erythromycin is unstable in gastric fluids. Particle size reduction will increase its surface area exposed to the gastric acid, thus increasing the extent of degradation and decreasing the drug absorption.

A32   E

Polymorphs may differ substantially in density, melting point, solubility and dissolution rates.

A33   E

All the listed factors can influence the bioavailability of drugs from hard gelatin capsules. Other factors that can influence the bioavailability of drugs from hard gelatin capsules include the salt form and crystal form of the drug; the chemical stability of the drug; and the nature and quantity of the diluents, lubricant and wetting agent.

A34   E

Hydroxypropylmethylcellulose is a water-soluble polymer and should have no significant effect on the rate of disintegration of the tablet core. Thus, it should not protect the drug from gastric acid, delay the rate of drug release or influence the rate of drug absorption.

A35   A

The saturation solubility of the drug in solution is a property of the compound in a specific dissolving medium, and it can be increased only by increasing the temperature.

A36   E

Usually the total plasma drug concentration is measured, as it is difficult to separate protein-bound drug from unbound drug in routine clinical practice. The protein binding reported in the clinical literature is used to calculate the free drug concentration in plasma.

A37   B

k (elimination rate constant) $= 0.693/\text{half-life} = 0.693/2.2 = 0.32\,h^{-1}$

A38   E

A distinct distribution phase is seen in the two-compartment model. On the other hand, only the elimination phase is seen in the one-compartment model.

A39   D

Usually it takes 4.5 distribution half-lives for a drug to be completely distributed. Therefore, $4.5 \times 35\,min = 158\,min = 2.6\,h$.

A40   B

Generally, the absorption rate constant is greater than the distribution rate constant, which is greater than the elimination rate constant for a two-compartment model drug.

A41   B

The profile shows extravascular (all routes of drug administration except intravenous administration involve an absorption phase) drug administration for a one-compartment model drug. It is characterised by absorption and elimination phases.

A42   D

The profile (straight lines) shows that the absorption and elimination follow first-order kinetics. Also note that the y-axis is in the log scale.

A43   A

The equation describes a one-compartment model of extravascular drug administration. C_0 is the initial drug concentration at time (t) zero, k_a is the absorption rate constant and k is the elimination rate constant.

A44   B

The slope of the first (solid) line gives the absorption rate constant. The slope is calculated by taking any two points on the line, i.e. $\ln y_2 - \ln y_1 / t_2 - t_1$. The rate constant is calculated by the method of residuals.

A45   A

The slope of the second (dotted) line gives the elimination rate constant. The slope is calculated by taking any two points on the line, i.e.

$\ln y_2 - \ln y_1 / t_2 - t_1$. The rate constant is calculated by extrapolating the terminal phase of the curve to the y-axis.

A46   E

Both the straight lines meet at the y-intercept at $x = 0$, i.e. initial drug concentration at $t = 0$.

A47   C

The AUC gives the extent of drug that is in systemic circulation. It is calculated by the trapezoidal method. The unit for AUC is $\mu g\, h/mL$.

A48   D

The half-life is independent of the route of drug administration. It is calculated from the terminal slope of the curve. The slope gives the elimination rate constant and $0.693/k$ gives the half-life of the drug. The only difference between extravascular and intravascular drug administration is the absorption phase. Once the drug is absorbed into the systemic circulation, the body handles the drug in the same way as intravascular drug administration.

A49   C

CL_{tot} is a summation of hepatic clearance (CL_h) and renal clearance (CL_r). These are the two main pathways of drug elimination. $CL_r = CL_{tot} - CL_h = 11.76 - 10 = 1.76\,L/h$. CL_r accounts for the drug that is excreted unchanged in the urine. The percentage of unchanged drug excreted in the urine is $100\,(1.76/11.76) = 15\%$.

A50   B

Generally, V_D gives an idea of the extent of drug distribution in the body and is calculated from single-dose intravenous administration. This route is used since 100% of the drug reaches the systemic circulation. The formula for calculating V_D is dose divided by initial drug concentration.

A51   A

Drugs bound strongly to plasma protein generally remain in the vascular compartment and therefore will have a low V_D. This is because the drug bound to plasma proteins can not be distributed to the tissue and only the unbound drug can cross the membrane to distribute to the tissues.

A52   E

There is an equilibrium between unbound and protein-bound drug in plasma. The fraction of the drug unbound in plasma increases when the binding of the drug to the protein-binding sites is inhibited by another drug that competes for the same binding site. Alternatively, if the drugs bind to different binding sties, the drug binding to one of the sites can alter the conformation of the protein and prevent the binding of drugs to other binding sites (allosteric inhibition). Similarly, when the protein concentration is decreased, the fraction unbound increases.

A53   D

Drugs that have a large V_D are generally bound strongly to tissues. As a result, the drug stays in the body for a longer time period. Similarly, the drug bound to plasma proteins will also have a long half-life, since the protein-bound drug can not be metabolised or excreted from the body.

A54   D

Generally, the absorption rate constant is greater than the elimination rate constant, but in the case of flip-flop pharmacokinetics the reverse is true. This happens if the drug is absorbed slowly. For example, transdermal drug application results in slower absorption since the drug has to cross several skin layers before reaching the blood. Since the first-order rate constant and half-life are inversely related, a decrease in the first-order rate constant will increase the half-life.

A55   B

In non-linear pharmacokinetics, the drug elimination becomes zero-order, i.e. a constant amount of drug is eliminated irrespective of the drug concentration. This happens when the enzyme-binding sites are saturated. Therefore, when a larger dose is administered, drug metabolism slows down and as a result the elimination, half-life and AUC are not proportional to dose. In other words, the half-life changes with dose.

A56   D

In non-linear pharmacokinetics, the drug elimination follows Michaelis–Menten kinetics. The Michaelis–Menten constant V_{max} (maximum rate of elimination, i.e. elimination when all the enzyme-binding sites are saturated; unit = amount/time) is used for dosing calculations. When the dosing rate becomes equal to or greater than V_{max}, the drug will follow non-linear pharmacokinetics. In this case, 300 mg twice daily (600 mg/day) and 150 mg three times daily (450 mg/day) exceed V_{max} (15 mg/h, i.e. 360 mg/day).

A57   C

The time to reach steady-state concentration (i.e. 95% of steady state) is 4.5 elimination half-lives. Steady state represents equilibrium between input (dosing rate) and output (drug clearance).

A58   A

Clearance is used for maintenance dose calculations. Maintenance dose = steady-state plasma concentration × clearance × dosing frequency.

A59   B

Loading dose = steady-state plasma concentration × volume of distribution.

A60   B

Time to reach steady state is 4.5 half-lives. Therefore, $4.5 \times 0.9 = 4$ h.

A61   A

Drug accumulation is calculated using the equation $1/(1-e^{-k\tau})$, i.e. the drug accumulation depends on the elimination rate constant (or half-life) and the dosing frequency. If the dosing frequency is much longer than the half-life of the drug, then minimal drug accumulation will be seen during multiple dosing.

A62   B

Since it is very close to the next dosing period, the dose can be skipped in order to avoid overdosing.

A63   B

If the dosing frequency is equal to the drug's half-life, then the loading dose is generally twice the maintenance dose.

A64   C

$CL_{tot} = V_D k$. Therefore, $k = CL_{tot}/V_D = 7.1/44.9 = 0.16\,h^{-1}$

A65   A

$C_{max-n} = dose/V_D[(1 - e^{-nk\tau})/(1 - e^{-k\tau})]$

where n = dose number = fourth dose; and τ = dosing frequency = 4 h.

$C_{max-4} = 200/44.9[(1 - e^{-4\times0.16\times4})/(1 - e^{-0.16\times4})] = 8.71\,mg/L$

A66   B

$C_{min-n} = C_{max-n}\, e^{-k\tau}$

where n = dose number = fourth dose; and τ = dosing frequency = 4 h.

$C_{min-4} = 8.71e^{-0.16\times4} = 4.60\,mg/L$

A67   E

Half-life = 0.693/0.16 = 4.33 h.
Time to reach steady state is 4.5 half-lives = 4.5 × 4.33 = 19.5 h. Steady state is reached between the fifth and sixth doses.

$C_{max-ss} = dose/[V_D(1 - e^{-k\tau})]$
$C_{min-ss} = 200/[44.9(1 - e^{-0.16\times4})] = 9.41\,mg/L$

A68   D

$C_{min-ss} = C_{max-ss}\, e^{-k\tau}$
$C_{min-ss} = 9.41\, e^{-0.16\times4} = 4.96\,mg/L$

A69   D

$C_{p-n} = dose/V_D[(1 - e^{-nk\tau})/(1 - e^{-k\tau})]e^{-kt}$

where n = dose number = third dose; τ = dosing frequency = 4 h; and t = 2 h.

$C_{p-3} = 200/44.9[(1 - e^{-3\times0.16\times4})/(1 - e^{-0.16\times4})]e^{-0.16\times2} = 5.87\,mg/L$

A70   C

$C_{p-n} = dose/V_D[(1 - e^{-nk\tau})/(1 - e^{-k\tau})]e^{-kt} - e^{-kt_{miss}}$

where n = dose number = third dose; τ = dosing frequency; t = 2 h; and t_{miss} = time since the dose was missed. The dosing frequency is 4 h, missing a dose will mean missing 4 h, and the time since the missed dose is 6 h (4 h missed from second dose plus 2 h from the third dose).

$$C_{p-3} = 200/44.9[(1 - e^{-3 \times 0.16 \times 4})/(1 - e^{-0.16 \times 4})]e^{-0.16 \times 2}$$
$$- e^{-0.16 \times 6} = 4.17 \text{ mg/L}$$

A71   D

$$C_{p-n} = D_0/V_D[(1 - e^{-nk\tau})/(1 - e^{-k\tau})]e^{-kt} - e^{-kt_{miss}} + e^{-kt_{actual}}$$

where n = dose number = fourth dose; τ = dosing frequency; and t = 2 h.
In this case, the late dose should be considered missed and subtracted from the equation. Then the term for the late administration should be added.
t_{miss} = time since the last dose was missed, i.e. since the dosing frequency is 4 h, missing a dose will mean missing $4 + 2$ h = 6 h (similar to the previous problem).
t_{actual} = time since the actual drug administration, i.e. second dose was given 1 h late, so the time since the actual administration is 3 h (from the second dose) plus 2 h (from the third dose) = 5 h.

$$C_{p-3} = 200/44.9[(1 - e^{-3 \times 0.16 \times 4})/(1 - e^{-0.16 \times 4})]e^{-0.16 \times 2}$$
$$- e^{-0.16 \times 6} + e^{-0.16 \times 5} = 6.17 \text{ mg/L}$$

A72   B

Oral extended-release systems are designed to release the drug more slowly than the absorption rate.

A73   B

Ideally, drugs with a half-life of 2–6 h are suitable for oral extended-release systems. Similarly, drugs that are uniformly absorbed throughout the gastro-intestinal tract are preferred.

A74   D

Enteric-coated systems are used for drugs that are unstable in the acidic pH of the stomach; they release the drug in the intestine, e.g. erythromycin enteric-coated tablets. In this case, the drug release is not extended but is delayed until it reaches the intestine. The system is designed to release the entire drug in the intestine.

A75   B

In the case of water-insoluble matrix tablets (e.g. wax or water-insoluble polymer), drug release follows the Higuchi equation:

$$Q = \left[\frac{D\varepsilon}{\tau}(2A - \varepsilon C_s)tC_s\right]^{1/2}$$

Here, the amount of drug (Q) depleted from the matrix per unit time t is dependent on the diffusion coefficient of the drug in the matrix (D), the porosity of the matrix (ε), the tortuosity (τ) of the capillaries in the matrix, the solubility or saturation concentration of the drug in the matrix (C_s) and the total concentration of drug in the matrix both dissolved and undissolved (A). Since the drug release is dependent on the porosity of the matrix and the tortuous diffusion pathway in the matrix, the drug release is non-linear.

A76   B

Drug release from polymeric systems is based on the water solubility of the polymer and diffusion of water into the delivery system. In water-soluble matrix systems, the polymer dissolves slowly from the surface and releases the drug. The size of the system also decreases with time. Surface eroding systems can achieve better control of drug release over bulk eroding systems.

A77   C

Osmotic controlled-release systems can achieve absolute zero-order release. The drug is released at a controlled rate based on the osmotic pressure in the system.

A78   C

Cationic exchange resins release the drug by exchanging the cationic drug with cations found in the gastrointestinal drug. In the stomach the drug is exchanged for hydrogen ions, and in the intestine the drug is exchanged for sodium ions.

A79   C

Entry of water into the system causes a difference in the osmotic pressure, and the drug released is dependent on the resultant osmotic pressure.

A80   C

The dose contained in a single-unit sustained-/controlled-release system is higher than in an immediate-release dosage form. Hence, there is a chance of dose dumping if the system fails. However, in the case of multiparticulate systems, the drug is encapsulated in individual micro-beads, and therefore there is relatively less chance of dose dumping.

A81   B

In a push–pull osmotic system, the drug is separated from the osmogen layer by a polymeric layer. The whole system is coated with a semipermeable polymer. Water enters through the semipermeable membrane and causes an increase in hydrostatic pressure in the osmogen layer. The osmogen layer expands and pushes the drug layer. The drug is then released through a laser-drilled hole in the system.

A82   A

The drug should have an aqueous solubility of at least 0.1 mg/mL in order to show a good release profile. Drug A has optimal water solubility. On the other hand, if the drug has very high aqueous water solubility (Drug D), it becomes difficult to control drug release.

A83   A

Drugs that are passively absorbed are preferred for oral sustained-/controlled-release systems. Unlike passive diffusion, active transport takes place only in certain regions of the gastrointestinal tract, such as the small intestine. The absorption rate should be greater than the drug release rate. Drugs should have an absorption rate constant of at least $0.17\,h^{-1}$. If the drug is absorbed slowly (Drug C), then it may not warrant an extended-release system. Similarly, if the drug absorption rate is very high (Drug D), then the dose becomes high for a sustained-/controlled-release system. The drug should have good oral bioavailability. Based on all these criteria, Drug A has the optimal absorption characteristics.

A84   D

If the drug has a very long half-life, then there may not be a need for sustained-/controlled-release system, since the drug is already long-acting. An ideal half-life is 4–6 h. If the half-life is very short, then a higher dose may be needed, which may be limited by the size of the sustained-/controlled-release system.

A85   C

If the dose of the drug is very high, then the size of the system becomes large. A desirable dose for a sustained-/controlled-release system is < 500 mg.

A86   A

Drug A has the optimal solubility, dose and absorption characteristics for developing a sustained-/controlled-release system. For the other drugs, one or more factors are not optimal for a sustained-/controlled-release system.

A87   C

Although the initial dose is required to be released rapidly, the maintenance dose is released slowly for a prolonged therapeutic effect. Modified-release formulations use a chemical or physical barrier to allow slow release of the drug. It is, however, the priming dose that should provide a rapid initial release of the drug.

A88   D

Maintenance of the therapeutic plasma concentration, improved compliance and the reduction of localised gastrointestinal effects are all advantages. Although these dosage forms are expensive to manufacture, there is a cost saving due to better disease state management. The fact that a larger amount of the drug than a single dose is incorporated has the potential for an overdose should the chemical or physical barrier be compromised.

A89   A

Drugs should ideally have a biological half-life of 4–6 h so that accumulation does not occur and significant amounts of pharmacologically inactive metabolites are not produced by first-pass metabolism. The dosage should also not exceed 325 mg. A factor that influences modified-release dosage forms is that the time the dosage form is resident in the gastrointestinal tract is influenced by both stomach emptying and intestinal transit time.

A90   E

The therapeutic index is not a physicochemical property of a drug. Although there may be safety implications for drugs with narrow therapeutic indices formulated as modified-release dosage forms, the therapeutic index is not a factor that influences the release of the drug from these delivery systems.

A91 E

In modified-release dosage forms, the device is hydrated and water diffuses into the dosage form. The drug should dissolve and then diffuse out of the device.

A92 E

The matrix and membrane formers, channelling agents and solubilisers all play a role in controlling the release of a drug from a modified-release dosage form. The lubricant is an excipient necessary for the manufacturing process and does not control the release of the drug from this delivery system.

A93 B

Carbopol is the hydrophilic excipient that forms the hydrophilic matrix in a modified-release dosage form. Methacrylate copolymer and polyvinyl acetate are examples of inert excipients, while carnauba wax and hydrogenated vegetable oils are lipid excipients.

A94 A

Although the polymer is inert, an essential characteristic is that it is insoluble in the gastrointestinal tract. The release rate of the drug depends on the drug diffusing through a network of capillaries and may be modified by changing the porosity of the matrix. There is concern about initiators and catalysts used in the preparation of the polymer of the matrix leaching out with the drug; since these matrices remain intact during gastrointestinal transit, this has led to concerns about impaction and patient concern when seeing the matrix as a 'ghost' in the stool.

A95 E

Channelling agents control drug release from the matrix and as such they are hydrophilic, are solid at room temperature, melt at body temperature and are soluble in the gastrointestinal tract. They enhance the diffusion of the drug. Examples include polyethylene glycols, polyols and surfactants. They do not improve the flow properties of formulations during tabletting, as this is the role of the lubricant. The role of channelling agents is to create capillaries through which the dissolved drug may diffuse to be released from then insoluble polymer and be water soluble; examples include sodium chloride and sugars.

A96   C

As the outer hydrated matrix of a hydrophilic colloid erodes, it becomes less and less concentrated, with the rate of erosion depending on the nature of the colloid. Diffusion of drug through the hydrated matrix controls its rate of release.

A97   B

The diffusion pathway is via the continuous phase in the interstices for true gels and not viscous matrices. In true gels the cross-links are fixed after the gel has formed, whereas there are no 'fixed' cross-links in viscous matrices. The bulk viscosity of the true gel is derived from the cross-linked polymer chains. Bulk viscosity does not correlate well with diffusion for true gels, but diffusion does correlate with microviscosity for these systems.

A98   C

An advantage of hydrophilic systems is that they can sustain high drug loading. Disadvantages include the release of the drug being dependent on two diffusion processes; erosion of the outer layer complicating release of the drug; the requirement of batch-to-batch consistency in the matrix-forming materials, other components and process parameters being difficult to achieve; and the need for optimal rate-controlling polymers for different active drugs.

A99   C

The gel modifiers associate with the matrix molecules and have a role to play in interactions at the molecular level, including cross-linking. They allow more complete, uniform and rapid hydration of the gel matrix; they modify the environment in the interstices of the gel either to speed up or slow down diffusion; and they suppress or promote the ionisation of ionisable polymers.

A100   A

The event that does not follow the immersion of a hydrophilic matrix containing a water-soluble drug in an aqueous medium is dissolution of the surface – the surface is in fact required to dissolve in order to provide a 'burst' effect.

Test 3

Particle science and calculations

Alison Haywood, Beverley D Glass, David S Jones

and Sanjay Garg

Solid-state properties	Powder flow
Particle size analysis	Granulation, drying
Size reduction	Coating and multiparticulates
Mixing	Pharmaceutical calculations

Introduction

This chapter includes important topics of pharmaceutics associated with powdered drugs, including solid-state properties, particle size analysis and reduction, mixing, powder flow, granulation, drying, and coating and multiparticulates. Since most active and inactive pharmaceutical ingredients occur in the solid state as amorphous powders or as crystals, there are often many formulation challenges (e.g. polymorphism, hydrate formation) that must be overcome in order to design dosage forms suitable for manufacture that guarantee their satisfactory performance in patients. As polymorphism may have an effect on the bioavailability of drugs, especially those with low aqueous solubility, it is important not only for pharmaceutical manufacturers but also for pharmacists and pharmacy students to be aware of factors that may cause a polymorphic change. Although this is largely the domain of the pharmaceutical scientist in the industry, there are certain conditions, such as high humidity during storage, that might bring about a change in the form of the drug and potentially its therapeutic efficacy. This chapter introduces solid-state properties, which, due to the fact that most drugs and excipients exist as solids, are of considerable pharmaceutical importance. A description follows of some of the macroscopic properties that are important in dosage form design, including particle size, analysis and reduction. Particle size control is an important aspect of dosage form design, beginning with the formulation and manufacturing stages right through to the efficacy of the drug product after administration to the patient. Particle size reduction is often required to facilitate efficient processing of drugs that need to be mixed or to ensure the production of an aesthetically pleasing suspension. Since very few drug products contain only one ingredient, mixing is required during manufacture to ensure the even distribution of the active pharmaceutical ingredient (API) through the dosage form. Effective mixing also results in a good appearance of the drug product and is required so that the API is released at the correct rate at the site of action. Mixing is required during the production of solid dosage forms, ranging from tablets through to dry powder inhalers. Since drug powders are used primarily to produce tablets and capsules, the flowability of these powders is of significance in the production of pharmaceutical dosage forms. After mixing, granulation, involving the formation of large multiparticles is undertaken to ensure even distribution of each ingredient throughout the mix. The drying process is often the last stage of manufacture before packaging and is important to ensure that the residual moisture is low enough so that the powder flows and the drug product does not deteriorate on storage. Coating tablets and capsules is undertaken for a number of reasons, but primarily to protect the API from the environment. Coating of pellets and beads, referred to as multiparticulates, is undertaken to produce drug products with an extended release; because this requires less frequent dosing by the patient, improved adherence is noted. The final section of this test is dedicated to pharmaceutical calculations, a critical role for the pharmacist in all areas of practice, whether in dispensing or manufacturing.

Questions

Q1   Paracetamol is known to exist in two polymorphic forms, namely Form 1 (monoclinic) and Form 2 (orthorhombic). Form 2 readily undergoes plastic deformation upon compaction in a tablet machine. The difference between the two forms of paracetamol is due to a difference in:

    A   ❏ crystal habit
    B   ❏ unit cells
    C   ❏ Miller indices
    D   ❏ hydrogen bonding
    E   ❏ van der Waals forces

Q2   Which one of the following statements is incorrect? A change in the polymorphic form of a drug can result in:

A   ❑ an increase in melting point
B   ❑ a decrease in solubility
C   ❑ an increase in solubility
D   ❑ a change in compression characteristics
E   ❑ the formation of a eutectic mixture

Q3   The anhydrous form of a drug:

A   ❑ contains a solvent of crystallisation
B   ❑ contains a hydrate of crystallisation
C   ❑ is less soluble than the crystal hydrate
D   ❑ is less soluble than the non-aqueous solvate
E   ❑ displays decreased absorption and bioavailability compared with the hydrated form

Q4   There are important formulation aspects related to the crystal habit of a solid drug. Which one of the following statements is incorrect?

A   ❑ needle-like crystals of a drug in suspension are easier to inject through a fine needle than plate-like crystals
B   ❑ the plate-like crystals of tolbutamide can cause powder bridging in the hopper of the tablet machine
C   ❑ the needle-like crystals of ibuprofen have poor powder flow properties
D   ❑ changes in crystal habit can cause changes to the strength and disintegration time of tablets
E   ❑ the ability of a suspension to form a 'cake' is influenced by the crystal habit of a drug

Q5   Factors contributing to a change in crystal habit during crystallisation include:

1   ❑ choice of solvent
2   ❑ presence of impurities
3   ❑ addition of surfactants
     A ❑ 1, 2 and 3
     B ❑ 1 and 2
     C ❑ 1 and 3
     D ❑ 1
     E ❑ 3

Q6   The crystal form of a drug may influence the following processes during the manufacture and testing of dosage forms except:

A   ❑ powder flow
B   ❑ compressibility
C   ❑ tablet coating
D   ❑ dissolution rate
E   ❑ filtration processes

Q7   A eutectic mixture of two substances results in:

A   ❑ an increase in the water of crystallisation
B   ❑ a formation of crystal solvates
C   ❑ a decrease in water solubility
D   ❑ a decrease in melting point
E   ❑ a decrease in bioavailability

Q8   The eutectic point of a mixture of two solids, A and B, is:

A   ❑ the point where A and B dissolve to form a solution
B   ❑ the minimum melting point of any possible combination of A and B
C   ❑ the maximum melting point of any possible combination of A and B
D   ❑ the point below which a liquid phase exists
E   ❑ the point where A and B exhibit limited water solubility

Questions 9 and 10 involve the following case:

A pharmacist wishes to determine the particle size of spironolactone for the preparation of an oral liquid suspension. Sieve analysis is a method commonly used in the determination of particle size of powders. The sieve diameter is the width of the minimum square aperture through which the particle passes.

Q9   Sieve analysis is:

A   ❑ a largely automated process
B   ❑ a rapid form of particle size analysis
C   ❑ a process that utilises a series or stack of sieves
D   ❑ used only for dry powders
E   ❑ able to measure particle diameters between 1 μm and 10 μm

Q10 The ISO range of analysis (in μm) of powders using sieves is between:

A ❑ 10 and 45
B ❑ 45 and 100
C ❑ 45 and 1000
D ❑ 100 and 450
E ❑ 450 and 1000

Q11 If a chemical substance is described in a pharmacopoeia as a 'very fine powder', the equivalent diameter of the particle size is such that the coarsest sieve diameter (in μm) is:

A ❑ 100
B ❑ 180
C ❑ 180, and the sieve diameter through which no more than 40% of powder must pass is 100 μm
D ❑ 100, and the sieve diameter through which no more than 40% of powder must pass is 80 μm
E ❑ 125

Q12 A particle of size 0.01 μm can be measured using:

A ❑ an environmental microscope
B ❑ a light microscope
C ❑ a Fraunhofer diffraction microscope
D ❑ a photon-correlation microscope
E ❑ a transmission electron microscope

Q13 The most important factor governing the method selection for determining particle size distribution of a powder is:

A ❑ speed of analysis
B ❑ particle size range
C ❑ cost
D ❑ environment that most closely resembles the conditions in which the powder will be processed or handled
E ❑ data-processing capability

Q14 The following method of particle size analysis is not suitable for particles over a wide range of diameters:

A ❑ laser light-scattering method
B ❑ Coulter counter method
C ❑ gravitational sedimentation method
D ❑ microscope method
E ❑ sieve method

Q15 Photon correlation spectroscopy (PCS):

A ❏ is dependent on the particle shape of a substance
B ❏ uses the principle of Brownian motion to measure particle size
C ❏ is not affected by the viscosity of the suspending fluid
D ❏ is not affected by temperature
E ❏ uses Stokes' equation as the basis for the calculation of particle diameter

Q16 An electric sensing zone method (Coulter counter) is not suitable for size analysis of aerosol particles because:

1 ❏ the particles would be affected by Brownian motion
2 ❏ Stokes' law allows only for terminal velocity conditions
3 ❏ the sample measurement environment has to be in a liquid medium

A ❏ 1, 2 and 3
B ❏ 1 and 2
C ❏ 2 and 3
D ❏ 1
E ❏ 3

Q17 The limitation(s) of using Stokes' equation for determining particle diameters is (are):

1 ❏ a near-spherical particle shape is assumed
2 ❏ no particle aggregation may be present
3 ❏ there is a low settling velocity

A ❏ 1, 2 and 3
B ❏ 1 and 2
C ❏ 2 and 3
D ❏ 1
E ❏ 3

Q18 Which of the following is not an important factor to consider when selecting an appropriate particle size separation method?

A ❏ elutriation index
B ❏ prevention of environmental pollution
C ❏ grade efficiency
D ❏ sharpness index
E ❏ pharmacopoeial requirements

Q19 Which of the following statements is true regarding fluid energy milling?

A ❏ provides a larger particle size reduction range than ball milling

B ❏ provides a larger particle size reduction range than pin milling

C ❏ utilises a high-pressure jet of air

D ❏ is commonly employed to produce coarse powders

E ❏ can result in particle size reduction due to particle agglomeration

Q20 Fluid energy milling utilises which of the following methods of particle size reduction?

A ❏ attrition

B ❏ compression

C ❏ impact

D ❏ compression and attrition

E ❏ impact and attrition

Q21 Which of the following statements regarding particle size reduction processes is (are) correct? Continued milling can cause particle size enlargement:

1 ❏ only in the case of particles with a diameter of less than 5 μm

2 ❏ because of the dominance of cohesive forces between particles over comminution forces

3 ❏ because comminution forces are distributed over a decreasing surface area

A ❏ 1, 2 and 3

B ❏ 1 and 2

C ❏ 2 and 3

D ❏ 1

E ❏ 3

Q22 Which one of the following statements is incorrect with reference to particle size reduction?

A ❏ particle size reduction causes changes in particle size distribution

B ❏ the extent of particle size reduction is always related to milling time

C ❏ most of the energy put into a comminution operation effects size reduction

D ❑ the lower particle size limit of a milling operation is dependent on the energy input and on material properties

E ❑ materials with a moisture content below 5% are suitable for dry grinding

Q23 Which of the following statements is incorrect in the case of milling drug powders?

A ❑ milling can result in degradation of thermolabile drugs
B ❑ milling can result in changes in polymorphic forms of a drug
C ❑ increased particle size results in a decreased surface area
D ❑ decreased surface area results in an increased dissolution rate
E ❑ milling can result in changes to the bioavailability of a drug

Q24 When considering particle size reduction of a soft waxy substance such as stearic acid, which of the following statements is (are) true?

1 ❑ stearic acid is capable of absorbing large amounts of energy through elastic and plastic deformation
2 ❑ stearic acid can be more easily reduced in size by lowering the temperature below the glass transition point of the material
3 ❑ stearic acid can be more easily reduced in size by decreasing the moisture content of the material

A ❑ 1, 2 and 3
B ❑ 1 and 2
C ❑ 2 and 3
D ❑ 1
E ❑ 3

Q25 Which of the following is not a benefit of a reduction in particle size?

A ❑ improved percutaneous absorption in ointment formulations
B ❑ improved function of the lubricant in tablets
C ❑ decreased sedimentation rate in suspension formulations
D ❑ decreased solubility of poorly soluble drugs in solution
E ❑ increased absorption and bioavailability of drugs that are poorly soluble at the gastrointestinal pH

Q26 The mixing mechanism in which groups of adjacent particles are transferred from one location in the powder mass to another is described as:

A ❑ positive mixing
B ❑ negative mixing

C   ❏ diffusive mixing
D   ❏ convective mixing
E   ❏ shear mixing

Q27 Which of the following statements is (are) true regarding powder segregation (demixing)? Powder segregation:

1   ❏ may occur due to vibration of the powder bed
2   ❏ is more likely to occur with particles that have greater flowability
3   ❏ may result in unacceptable weight variation of tablets
    A ❏ 1, 2 and 3
    B ❏ 1 and 2
    C ❏ 2 and 3
    D ❏ 1 and 3
    E ❏ 3

Q28 The process of powder segregation in which smaller particles fall through the voids between larger particles and move to the bottom of the powder mass is known as:

A   ❏ elutriation segregation
B   ❏ percolation segregation
C   ❏ trajectory segregation
D   ❏ fluidisation segregation
E   ❏ 'dusting out'

Q29 When compared with a spherical particle, a non-spherical particle:

A   ❏ exhibits increased flowability
B   ❏ will segregate more easily
C   ❏ will decrease the likelihood of 'dusting out'
D   ❏ has a decreased surface area to weight ratio
E   ❏ is more likely to have cohesive effects with other particles

Q30 Which of the following statements is true? The ordered mixing of powders:

A   ❏ has a tendency to increase segregation
B   ❏ decreases flow properties
C   ❏ is not possible if the mix is exposed to excessive vibration
D   ❏ is independent of the amount of carrier particles available
E   ❏ is independent of the particle size range of carrier particles

Q31 Which of the following statements is incorrect? Diffusion mixing:

A   ❏ is generally preferred for mixes containing potent drugs

B   ❏ is inefficient in the case of overfilling, since sufficient bed dilation may not occur

C   ❏ is able to produce a random mix

D   ❏ utilises a high speed of mixing

E   ❏ is preferable if high shear is needed to break up aggregates of cohered material

Q32 Which of the following statements is (are) true? Tumbling mixers:

1   ❏ may cause segregation if there are significant differences in particle size

2   ❏ are useful for poorly flowing or cohesive powders

3   ❏ are commonly used in the blending of lubricants and glidants with granules before tabletting

A ❏ 1, 2 and 3

B ❏ 1 and 2

C ❏ 2 and 3

D ❏ 1 and 3

E ❏ 3

Q33 Which of the following statements is (are) true? A ribbon mixer:

A   ❏ is more likely to cause segregation than a tumbling mixer

B   ❏ is inefficient in mixing poorly flowing powders

C   ❏ is easy to clean

D   ❏ eliminates 'dead spots'

E   ❏ utilises high shear to break up drug aggregates

Questions 34–36 involve the following case:

You are required to formulate a 60 mg tablet containing 50 µg of a potent active ingredient. You are requested to estimate the particle size required to produce a uniform tablet.

Q34 When calculating the particle size required, which of the following statements is true?

A   ❏ the results are dependent on changes in the particle size range

B   ❏ the results are independent of particle shape

C ❏ the results are independent of whether a random mix is achieved

D ❏ it is necessary only to calculate the particle size of the excipients

E ❏ it is necessary only to calculate the particle size of the active ingredient

Q35 Your calculations suggest that in order to meet product specification, the particle size of the components should be less than 26 µm. What suggestion would you make?

A ❏ increase the mixing time
B ❏ decrease the tablet weight
C ❏ decrease the content variation by decreasing the scale of scrutiny
D ❏ decrease the number of particles in the scale of scrutiny by decreasing the particle size
E ❏ decrease the particle size further

Q36 Which of the following statements is incorrect? Powders with a small particle size:

A ❏ may result in material adhering to machine surfaces
B ❏ should be mixed at a relative humidity of less than 40%
C ❏ tend to be cohesive
D ❏ exhibit poor flow properties
E ❏ will decrease the number of particles in the scale of scrutiny, thereby reducing variation in content

Q37 Mixing of semisolids:

A ❏ is not possible with planetary mixers
B ❏ results in a lower frequency of 'dead spots' than mixing powders
C ❏ relies on convective mixing processes
D ❏ relies on diffusion mixing processes
E ❏ requires a small clearance between the blades/paddles and the mixing vessel

Q38 With reference to powder flow, cohesion:

A ❏ increases as particle size increases
B ❏ can occur between particles and machine components (e.g. hoppers)

C ❑ is composed mainly of short-range van der Waals forces

D ❑ is independent of electrostatic forces

E ❑ is independent of relative humidity

Q39 In the case of the packing geometry of powders:

A ❑ the true density of a powder is less than the bulk density

B ❑ an increase in the number of interparticulate contacts causes a decrease in cohesion

C ❑ an increase in the number of interparticulate contacts causes an increase in porosity

D ❑ a decrease in particle size range will result in a more closely packed cohesive powder

E ❑ porosity can be misleading if the interparticulate pore size distributions are not known

Q40 Powders that exhibit good flow properties have:

A ❑ a high angle of repose

B ❑ more than one angle of repose

C ❑ a small particle size

D ❑ a high surface area to weight ratio

E ❑ a high particle density

Q41 Arch or bridge formation is more likely to occur during powder packing with powders that have:

A ❑ irregular shapes

B ❑ a small particle size range

C ❑ decreased electrostatic forces

D ❑ a large particle size

E ❑ decreased cohesive forces

Q42 Which of the following statements is (are) correct? Two different powders are likely to exhibit similar packing geometries if they have the same:

1 ❑ interparticulate pore size distributions

2 ❑ porosity

3 ❑ particle size

A ❑ 1, 2 and 3

B ❑ 1 and 3

C ❑ 2 and 3

D ❑ 1 and 2

E ❑ 3

Q43 In the case of powder flow through the orifice of a hopper, which of the following statements is true?

A ❏ when the powder head is much greater than the orifice diameter, flow rate is independent of the powder head

B ❏ the wider the hopper, the more likely that 'arching' will occur

C ❏ powders with low wall friction angles will empty slowly

D ❏ particle segregation is reduced in square-bottomed hoppers if the hopper is partially emptied before refilling

E ❏ steep-walled hoppers are likely to cause funnel flow ('rat-holing')

Q44 Powder flow is influenced by:

1 ❏ design of the hopper
2 ❏ angle of repose
3 ❏ bulk density
A ❏ 1, 2 and 3
B ❏ 1 and 3
C ❏ 2 and 3
D ❏ 1 and 2
E ❏ 3

Q45 A parameter used as a direct method for characterising powder flow is:

A ❏ angle of repose
B ❏ Carr's index
C ❏ Hausner ratio
D ❏ hopper flow rate
E ❏ bulk density

Questions 46 and 47 relate to the following case:

A pharmaceutical manufacturer has found that the powder from one of the tablet formulations is not flowing reproducibly into the tablet die. They have asked you for suggestions to improve the situation.

Q46 Powder flow properties can be improved by:

A ❏ purchasing a finer grade of powder
B ❏ decreasing the relative humidity (RH) to below 20% in the tabletting area

C ❏ incorporating glidants such as hydroxypropylmethylcellulose (HPMC)

D ❏ using spray-dried excipients

E ❏ increasing the hopper wall angle

Q47 You would recommend the use of vibration-assisted hoppers if:

A ❏ the formulation can be altered

B ❏ the hopper orifice diameter can be adjusted

C ❏ the hopper width can be adjusted

D ❏ the hopper wall angle can be adjusted

E ❏ the powder arch strength within the hopper is greater than the stresses due to gravitational effects

Q48 Which of the following statements with regard to improving powder flowability is (are) true?

1 ❏ force feeders act by preventing powder arching over tablet dies

2 ❏ increasing the length of powder transfer pipes can reduce electrostatic charges and improve powder flow

3 ❏ an increase in moisture content can increase bulk density and reduce porosity of powders

A ❏ 1, 2 and 3

B ❏ 1 and 3

C ❏ 2 and 3

D ❏ 1 and 2

E ❏ 3

Q49 Which of the following statements is incorrect? Glidants improve flowability of powders by:

A ❏ reducing adhesion

B ❏ reducing cohesion

C ❏ increasing the angle of repose

D ❏ disrupting the continuous film of adsorbed water surrounding moist particles

E ❏ reducing bulk density of tightly packed powders

Q50 Weight uniformity may be improved by modifying powder flowability. However, in extreme cases, it may lead to a decrease in:

A ❏ particle size range

B ❏ content uniformity

C   ❏ powder segregation
D   ❏ particle size
E   ❏ costs

Q51   A large particle size distribution of granules may result in:

1   ❏ segregation of granules
2   ❏ tablets/capsules with large weight variations since machines fill by weight rather than volume
3   ❏ unacceptable drug content uniformity in the tablets/ capsules despite even distribution of drug within the granules
A   ❏ 1, 2 and 3
B   ❏ 1 and 3
C   ❏ 2 and 3
D   ❏ 1 and 2
E   ❏ 3

Q52   Granulation is normally a part of manufacturing tablets because:

A   ❏ demixing is not possible with granules
B   ❏ segregation is not possible with granules
C   ❏ a large variation in particle size distribution is not possible with granules
D   ❏ it is an inexpensive process
E   ❏ granules occupy less volume per unit weight, which is convenient for storage

Q53   Which of the following statements is true? In the case of wet granulation:

A   ❏ common granulating fluids include water, ethanol and propylene glycol, either alone or in combination
B   ❏ binding agents are seldom used in granulating fluids
C   ❏ the use of organic solvents as granulating fluids is not a suitable alternative to dry granulation for drugs sensitive to hydrolysis
D   ❏ the primary advantage of water is that it is non-flammable, and expensive safety precautions including the use of flame-proof equipment are not required
E   ❏ granules prepared by wet massing have a greater porosity than fluidised-bed granules

Q54 Adhesion and cohesion forces in liquid films between primary powder particles:

A ❑ are inversely proportional to the strength of the van der Waals forces of attraction

B ❑ decrease with an increase in contact area between particles

C ❑ increase with a decrease in interparticulate distance

D ❑ are due largely to thin, mobile films

E ❑ are unlikely to contribute significantly to final granule strength in dry granulation

Q55 Continued kneading/mixing during wet granulation will result in:

A ❑ a decrease in pore volume occupied by air

B ❑ a decrease in the density of the wet mass

C ❑ an increase in interparticulate distance

D ❑ a decrease in granule tensile strength

E ❑ an increase in the formation of the pendular state

Q56 The capillary state achieved during wet granulation is:

A ❑ an intermediate state between the pendular and funicular states

B ❑ due to interfacial forces in immobile liquid films between particles

C ❑ due to van der Waals forces between particles

D ❑ due to forces at the liquid–air interface

E ❑ achieved by increasing the separation of particles

Q57 Fluidised-bed granulation:

A ❑ is an efficient mixing process

B ❑ is capable of rapidly transforming usable granules into an unusable, overmassed system

C ❑ is not affected by the temperature of the granulating solution

D ❑ is not affected by humidity

E ❑ results in increased labour costs

Q58 The disadvantage(s) of fluidised-bed granulators compared with traditional high-speed mixer-granulators is (are):

1 ❑ transfer losses between equipment occur more frequently

2 ❑ overmassing is likely

3   ❏ optimisation requires extensive developmental work, not only during the initial formulation but also during scale-up

A ❏ 1, 2 and 3
B ❏ 1 and 3
C ❏ 2 and 3
D ❏ 1 and 2
E ❏ 3

Q59 Which of the following statements is incorrect? Extrusion and spheronisation:

A   ❏ increases bulk density
B   ❏ improves flowability of particles
C   ❏ is a less labour-intensive process than other forms of granulation
D   ❏ requires fewer excipients than other forms of granulation
E   ❏ reduces dust contamination

Q60 When comparing extrusion with conventional granulation by wet massing processes, which of the following statements is (are) true?

1   ❏ the extrudable wet mass needs to be wetter than that appropriate for conventional granulation
2   ❏ both processes are suitable for controlled-drug-release applications
3   ❏ both processes rely on a uniformly dispersed granulating fluid

A ❏ 1, 2 and 3
B ❏ 1 and 3
C ❏ 2 and 3
D ❏ 1
E ❏ 3

Q61 Solute migration during the drying stage of an extrusion process will result in:

A   ❏ a softer pellet being formed
B   ❏ modified surfaces, which may reduce adhesion of film coats
C   ❏ a decrease in the initial dissolution rate
D   ❏ a transition from cylindrical particles into 'dumbbells'
E   ❏ a decrease in the plasticity and cohesiveness of the wet mass

Q62 Which of the following statements is incorrect? 'Overdrying' a wet solid will result in:

A ❑ an increased potential of reaching the dew point with subsequent condensation

B ❑ an increase in the equilibrium moisture content of the solid

C ❑ a decrease in flow properties due to static charges

D ❑ an increase in costs, since energy is required in the drying process

E ❑ a decrease in compaction properties

Questions 63 and 64 involve the following case:

A pharmaceutical manufacturer requires a solution containing a thermolabile material to be dried.

Q63 Which of the following statements is incorrect when comparing spray drying and freeze drying?

A ❑ both processes result in an increase in bulk density

B ❑ both processes can be used for thermolabile materials

C ❑ since water is removed during both processes, the product can be packaged and reconstituted with water at a later stage

D ❑ both processes result in a product with an improved dissolution rate

E ❑ spray drying is a much faster process than freeze drying

Q64 Which of the following is a limitation of freeze drying?

A ❑ the potential for hydrolysis is increased

B ❑ the potential for oxidation is increased

C ❑ products formed are hygroscopic in nature

D ❑ the final product has decreased solubility

E ❑ salts may concentrate in the wet state and denature proteins

Q65 Fluidised-bed drying:

A ❑ uses conduction as a heat transfer method for drying wet solids

B ❑ requires a long processing time

C ❑ can cause aggregation of particles

D ❑ is unlikely to produce too much dust

E ❑ can result in explosions if adequate electrical earthing is not in place

Q66 Which of the following statements is correct? Microwave drying:

A ❑ utilises convection and conduction as a method of heat transfer

B ❑ is efficient at a wavelength range of 10 cm to 1 cm

C ❑ avoids problems of dust and attrition since the bed is stationary

D ❑ should never be used under vacuum

E ❑ can produce similar batch sizes commercially compared with fluidised-bed drying

Q67 Which of the following statements is true with reference to solute migration during drying?

A ❑ fluidised-bed drying is less likely to cause intragranular migration than microwave drying

B ❑ fluidised-bed drying is less likely to cause intergranular migration than static convective drying

C ❑ increasing the granule size will decrease solute migration

D ❑ decreasing the viscosity of the granulating fluid will decrease solute migration

E ❑ the use of aluminium lakes will increase mottling

Q68 Which of the following statements describing granulation is correct?

A ❑ it increases the moisture absorption and improves the flow

B ❑ it reduces the moisture absorption and decreases the density of granules

C ❑ it reduces the moisture absorption and increases the granule density

D ❑ it decreases the flow and the density of granules

E ❑ it has no effect on the density of powders

Q69 What is the correct order of the extrusion and spheronisation process?

A ❑ wet massing > extrusion > spheronisation > drying > screening

B ❑ extrusion > wet massing > spheronisation > drying > screening

C ❑ wet massing > extrusion > spheronisation > screening > drying

D   ❏ wet massing > extrusion > spheronisation > drying > dry mixing

E   ❏ wet massing > extrusion > screening > spheronisation > drying

Q70 Which of the following is true regarding reasons for granulation?

A   ❏ to prevent segregation of the constituents of the powder mix due to variations in the size and density

B   ❏ to improve flow, because poor flow can result in variations in tablet or capsule weights and composition

C   ❏ to improve compaction characteristics of the mix

D   ❏ to reduce dust and moisture absorption

E   ❏ all of the above

Q71 Which of the following is true regarding reasons for coating pharmaceutical products?

1   ❏ it enhances aesthetic appeal and brand image and helps in masking unpleasant tastes, colours and odours

2   ❏ it improves compliance by enabling the product to be easily swallowed by the patient

3   ❏ it can improve product stability by protecting against deterioration by environmental factors such as sunlight, temperature variations, moisture and environmental gases

A ❏ 1
B ❏ 2
C ❏ 1, 2 and 3
D ❏ 3
E ❏ 1 and 3

Q72 Which of the following statements is not true with reference to sugar coating of tablets?

A   ❏ it is one of the oldest pharmaceutical processes still in existence

B   ❏ it uses inexpensive raw materials

C   ❏ sugar-coated tablets are widely accepted by consumers

D   ❏ it can be completed within 2 h with modern equipment

E   ❏ it leads to addition of significant weight, which can be as much as double the weight of the uncoated tablets

Q73 Which of the following is the correct sequence of steps in sugar coating?

A ❏ sealing > subcoating > grossing > colour coating > polishing
B ❏ sealing > grossing > subcoating > colour coating > polishing
C ❏ sealing > subcoating > colour coating > grossing > polishing
D ❏ sealing > subcoating > colour coating > polishing > grossing
E ❏ subcoating > sealing > grossing > colour coating > polishing

Q74 Sealing in sugar coating offers initial protection and prevents tablet core ingredients from migrating into the coating and spoiling the appearance of the final product. Which of the following excipients is not used in sealing?

A ❏ shellac
B ❏ zein
C ❏ hydroxypropylmethylcellulose
D ❏ potassium chloride
E ❏ cellulose acetate phthalate

Q75 Which of the following is (are) critical variable(s) in sugar coating?

A ❏ tablet load size
B ❏ pan RPM (rotations per minute)
C ❏ quantity and temperature of drying and exhaust air
D ❏ spray rate, pattern, distance and atomisation pressure
E ❏ all of the above

Q76 Which of the following is the least desirable property with reference to coating of tablets?

A ❏ low friability
B ❏ high hardness
C ❏ pentagonal shape
D ❏ good surface adhesion
E ❏ light blue colour

Q77 Which of the following is incorrect regarding film coating?

A ❏ it requires a shorter time than sugar coating, and the process can be automated easily
B ❏ it is less complex than sugar coating
C ❏ it requires sealing of the tablets before coating
D ❏ use of solvents and effects on safety of workers is an issue
E ❏ the process adds only 10–20 µg to each tablet

Q78 Which of the following is (are) excipients used in film coating?

1 ❑ polymer(s)
2 ❑ solvent(s)
3 ❑ plasticiser(s)
4 ❑ optional ingredients such as colorants, opaquants and antimicrobial agents

A ❑ 1, 2 and 3
B ❑ 1, 2 and 4
C ❑ 1, 3 and 4
D ❑ 1, 2, 3 and 4
E ❑ 2, 3 and 4

Questions 79 and 80 involve the following case:

A formulation scientist has been asked to develop a controlled-release oral formulation for an active ingredient that is unstable at a pH below 5.0.

Q79 The best formulation option for this drug is:

A ❑ tablets with a very high degree of hardness
B ❑ beads loaded in capsules
C ❑ normal capsules with a slow-release matrix
D ❑ enteric-coated tablets with a core containing a slow-release matrix
E ❑ normal tablets containing a slow-release matrix

Q80 Which of the following is not a good choice for coating slow-release tablets?

A ❑ cellulose
B ❑ methacrylate
C ❑ polyvinyl alcohol
D ❑ shellac
E ❑ cellulose and methacrylate

Q81 Which of the following is a correct option regarding solvents used in coating?

1 ❑ the solvent should dissolve/disperse the polymer system
2 ❑ the solvent should be non-toxic and inert

3   ❏ the solvent should be environmentally friendly (there are limits on the acceptable amounts for residual solvents in pharmaceuticals for the safety of the patient)

A ❏ 1
B ❏ 2
C ❏ 3
D ❏ 1, 2 and 3
E ❏ 1 and 3

Q82 Which of the following is not true with reference to coating?

A   ❏ most of the coating polymers are available as individual powders, but many are also available as premixed powder blends and aqueous dispersions

B   ❏ only certain types of capsule can be coated

C   ❏ dry powder concentrates for reconstitution combine polymer, plasticiser and pigment in a single system for either aqueous or organic coating, which reduces inventory handling and raw material testing

D   ❏ aqueous-based coating systems, including aqueous dispersions, are gaining popularity, as they are solvent-free and environmentally friendly

E   ❏ gelatin capsule shells possess the problems of poor polymer adhesion, leeching out to the coating agent, and brittleness

Q83 You have been asked to prepare 50 mL of a 10% w/w solution of a drug. The quantity of drug in the formulation is:

A   ❏ 10.0 g
B   ❏ 0.1 kg
C   ❏ 5.0 g
D   ❏ 0.05 kg
E   ❏ 0.5 kg

Q84 The concentration of a drug in plasma is 25 ng/mL. This concentration expressed as % w/v is equivalent to:

A   ❏ 2.5
B   ❏ 0.025
C   ❏ 2.5×10^{-6}
D   ❏ 2.5×10^{-9}
E   ❏ 25×10^{-6}

Q85 You have been asked to prepare 100 mL of a 70% alcohol solution, starting from a 90% alcohol solution. The volume (in mL) of 90% alcohol required is:

A   ❑ 70.0
B   ❑ 77.8
C   ❑ 63.0
D   ❑ 67.0
E   ❑ 75.0

Q86 A solution of potassium permanganate has been provided at a concentration of 1 ppm. The quantity of potassium permanganate contained in 100 mL of this solution is:

A   ❑ 1 g
B   ❑ 1 mg
C   ❑ 100 µg
D   ❑ 1 µg
E   ❑ 10 µg

Q87 The quantity of a drug in 25 g of a cream is 1 g. The concentration of drug in the cream expressed as % w/w is:

A   ❑ 4.0
B   ❑ 0.04
C   ❑ 0.4
D   ❑ 1.0
E   ❑ 2.5

Q88 Oxytetracycline hydrochloride ($C_{22}H_{24}N_2O_9$.HCl; molecular weight 496) has the following structure:

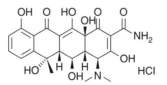

You have been asked to prepare 100 mL of a 0.2 M solution of oxytetracycline hydrochloride. The quantity of drug required is:

A   ❑ 99.2 g
B   ❑ 9.9 g
C   ❑ 4.96 g
D   ❑ 10 g
E   ❑ 0.99 g

Q89 You have been asked to manufacture a 2000 L batch of a drug at 10% w/v concentration. The quantity of drug required is:

A ☐ 200 kg
B ☐ 20 kg
C ☐ 2000 g
D ☐ 100 kg
E ☐ 1000 kg

Q90 A 100 L batch of an injection formulation containing 10% w/v of the drug was manufactured. During analysis, the drug concentration was found to be below specification (9.15% w/v). You have been asked to prepare a small batch of the formulation (10 L) containing excess drug, which will be added to the 100 L batch to ensure drug concentration at 10% w/v. The concentration of drug (% w/v) in the 10 L batch required to produce an overall 10% w/v concentration is:

A ☐ 0.85
B ☐ 11.0
C ☐ 10.0
D ☐ 18.5
E ☐ 1.85

Q91 A 200 L batch of an injection formulation containing 10% w/v of the drug was manufactured. During analysis, the drug concentration was found to be above specification (11.25% w/v). The volume of diluent required to produce 10% w/v concentration in the final batch is:

A ☐ 11.25 L
B ☐ 10.0 L
C ☐ 12.25 L
D ☐ 25.0 L
E ☐ 2.5 L

Q92 A vitamin formulation contains α-tocopheryl acetate (structure below). The molecular formula is $C_{31}H_{52}O_3$ and the molecular weight is 473.

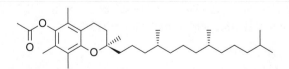

A company that manufactures a topical formulation of this drug claims that the nominal concentration of tocopheryl acetate is

5% w/w; however, the purity of the active substance is only 92%. The quantity of active agent required to manufacture a 200 kg batch is:

A ❑ 10 kg
B ❑ 5 kg
C ❑ 10.87 kg
D ❑ 9.2 kg
E ❑ 92.0 kg

Q93 An antibiotic solution contains 10% w/v oxytetracycline (molecular weight 460). In the manufacture of the formulation, oxytetracycline hydrochloride (molecular weight 496) is used with a purity of 91%. The quantity of oxytetracycline hydrochloride required to manufacture a 1 L batch of formulation containing 10% w/v oxytetracycline is:

A ❑ 100.0 g
B ❑ 107.82 g
C ❑ 118.5 g
D ❑ 9.27 g
E ❑ 92.7 g

Q94 You have been asked to prepare a 500 mL solution of a disinfectant that is to be diluted 1 in 10 in order to produce a solution with a final concentration of one part per 5000. The quantity (in g) of disinfectant in the starting solution is:

A ❑ 0.1
B ❑ 10.0
C ❑ 1.0
D ❑ 0.01
E ❑ 100.0

Q95 Calculate the quantity (in g) of a drug contained in 200 mL of a solution at a concentration of 1 in 5000:

A ❑ 0.04
B ❑ 0.4
C ❑ 0.01
D ❑ 0.1
E ❑ 4.0

Q96 The dose of drug to be administered to a patient by infusion is 10 µg/kg per hour. If the weight of the patient is 50 kg, calculate the total quantity of drug delivered to the patient in a 24-h period.

A ❑ 12 g
B ❑ 12 mg
C ❑ 120 mg
D ❑ 120 µg
E ❑ 1200 µg

Q97 A hospital pharmacist has been asked to prepare 100 g of an antibiotic cream with 0.05% w/w drug. The cream must be prepared by diluting a commercially available preparation containing 0.2% w/w of the drug. What quantity (in g) of the commercially available formulation is required to fill this prescription?

A ❑ 40.0
B ❑ 4.0
C ❑ 20.0
D ❑ 25.0
E ❑ 2.5

Q98 What is the concentration (% w/v) in the solution when 500 mg of a disinfectant is dissolved in 5 L of water?

A ❑ 0.1
B ❑ 1.0
C ❑ 0.01
D ❑ 0.001
E ❑ 10.0

Questions 99–100 involve the following case:

The formulation for an aqueous cream is as follows:	
Drug	5% w/w
Preservative	0.3% w/w
Surfactants	0.2% w/w
Base	qs 100% w/w

Q99 The quantity (in kg) of drug required to produce a 500 kg batch of this cream is:

A ❏ 25.0
B ❏ 2.5
C ❏ 50.0
D ❏ 0.25
E ❏ 0.5

Q100 The quantity (in kg) of preservative required to produce a 500 kg batch of this cream is:

A ❏ 0.15
B ❏ 1.5
C ❏ 15.0
D ❏ 0.75
E ❏ 7.5

Answers

A1   A

Materials in the solid state can be crystalline or amorphous. In crystalline materials, the molecules are packed in a defined order. The crystal habit describes the overall shape of a crystal (i.e. the polymorphic forms have the same chemical structure but different physical characteristics).

A2   E

Different polymorphic forms of a drug may have different melting points, solubilities and compression characteristics; they do not form eutectic mixtures.

A3   D

Drugs can entrap solvent or water when they crystallise (known as crystal solvates or crystal hydrates, respectively). Alternatively, no water may be entrapped during crystallisation (known as the anhydrous form). The anhydrous form is generally more soluble than the hydrate of the same drug, but less soluble than non-aqueous solvates.

A4   A

Plate-like crystals of a drug in suspension are easier to inject through a fine needle than needle-like crystals.

A5   A

Changes in the conditions of crystallisation of a solid, such as the choice of solvent, temperature, presence of impurities or addition of surfactants, will result in different crystal habits.

A6   C

A change in crystal form of a drug may cause changes in powder flow, compressibility, dissolution rates and filtration processes. It will not affect coating.

A7   D

A eutectic mixture is where two pure solid substances are mixed together in such proportions that the melting point is lower than each of the individual solid substances.

A8   B

The eutectic point of a mixture of solid substances is the temperature and composition at which the two solids crystallise out of solution. The melting point of this mixture is lower than each of the melting points of the two original solids.

A9   C

Sieve analysis is largely a non-automated process that utilises a series or stack of sieves. It analyses particles in the size range 5–125 000 µm. It is usually carried out on dry powders, although wet sieving for powders in liquid suspension is possible.

A10   C

The ISO range of analysis for sieves is between 45 µm and 1000 µm. Powders are usually defined as having a maximum particle size of 1000 µm.

A11   E

Standards for powders used pharmaceutically are provided in pharmacopoeias, which indicate that 'the degree of coarseness or fineness of a powder is expressed by reference to the nominal mesh aperture of the sieves used'. The coarsest sieve diameter for a very fine powder is 125 µm. It is important to always refer to particle sizes according to the appropriate equivalent diameters (expressed in mm, µm or nm); this should not be confused with the term 'sieve number', which has been used in some cases as a method of quantifying particle size.

A12   E

The transmission electron microscope (TEM) permits the lower particle size limit to be greatly extended over that possible with a light microscope. The TEM has a range of analysis limit of 0.001 µm.

A13   D

It is most appropriate to determine the particle size distribution of a powder in an environment that most closely resembles the conditions in which the powder will be processed or handled.

A14   C

The range of analysis for gravitational sedimentation methods is limited to particles with a diameter of between 5 μm and approximately 250 μm. Settling becomes prolonged with particles with a diameter below 5 μm and is subject to interference from convection, diffusion and Brownian motion.

A15   B

PCS is a laser light-scattering method of particle size analysis that uses the principle of Brownian motion to measure particle size. It is independent of external variations, except viscosity and temperature of the suspending fluid. Since it randomises particle orientations, any effects of particle shape are minimised.

A16   E

The electric stream sensing zone method or Coulter counter method requires powder samples to be dispersed in an electrolyte to form a very dilute suspension. This does not resemble the conditions under which the powder will be handled in vivo and would therefore not be the most appropriate method for particle size analysis.

A17   A

Stokes' equation for determining particle diameter in sedimentation methods of particle size analysis is based on the assumptions that the particles are near-spherical, the particles have a low settling velocity so that inertia is negligible, and there is no particle aggregation.

A18   A

The most efficient method of particle size separation should be selected based on pharmacopoeial requirements and particle properties. 'Grade efficiency' generally refers to the separation efficiency of the method and the 'sharpness index' can be used to quantify the sharpness of cut-off in a given size range. The potential for environmental pollution should also be considered.

A19   C

Fluid-energy milling utilises a high-pressure jet of fluid, usually air, to reduce particle size by particle impaction and attrition.

A20   E

Fluid-energy milling is a size-reduction method that acts by particle impaction and attrition.

A21   B

With continued milling, the interactive cohesive forces between small particles (diameters $< 5\,\mu m$) generally predominate over comminution stresses as the comminution forces are distributed over increasing surface areas. This results in particle agglomeration as opposed to particle fracture.

A22   C

Only a very small amount of the energy put into a comminution operation actually effects size reduction. The remainder is lost to elastic and plastic deformation of particles, deformation of metal machine parts, interparticulate friction, particle–machine wall friction, heat, sound and vibration.

A23   D

Size reduction results in decreased particle size, increased surface area and increased dissolution rate, which can have a marked effect on bioavailability.

A24   B

Soft substances are able to absorb large amounts of energy through elastic and plastic deformation and are therefore resistant to brittle fracture (crack initiation and propagation) and size reduction. By lowering the temperature at which milling occurs to below the glass transition temperature of the material, the material will undergo a transition from plastic to brittle behaviour and crack propagation is facilitated. Lowering the moisture content of the material will not be effective, since these substances soften when heated due to the milling action.

A25   D

Particle size reduction will result in a larger surface area of the particle being more readily dissolved in solution and having increased bioavailability when absorbed percutaneously or in the gastrointestinal tract. Lubricants with a small particle size are able to coat the granules or powders more effectively. Decreased particle size will result in a decreased sedimentation rate, which is a desirable property for suspension formulations.

A26 D

Diffusive mixing is where particles are distributed over a freshly developing surface, and shear mixing is when a 'layer' of material flows over another 'layer' (i.e. slip planes are set up within the mass). Positive mixing applies to systems that would spontaneously and completely mix over time, whereas in negative mixing the phases would separate in the absence of continuous agitation.

A27 A

Segregation can occur during the handling processes after powders have been satisfactorily mixed (e.g. during transfer to filling machines or in the hopper), resulting in unacceptable content and weight variation. Demixing is more likely to occur if the powder bed is subject to vibration and when the particles have greater flowability.

A28 B

Trajectory segregation is where large particles move greater distances than smaller particles before coming to rest due to their larger mass and kinetic energy. Elutriation segregation (also referred to as fluidisation segregation and 'dusting out') is when very small particles are blown upwards by turbulent air currents during mixing or when material is discharged from a container. When mixing is stopped, these particles will sediment and subsequently form a layer on top of the coarser particles.

A29 E

Spherical particles exhibit the greatest flowability, are more easily mixed and segregate more easily. Non-spherical particles have a greater surface area to weight ratio, which tends to decrease segregation by increasing any cohesive effects but will increase the likelihood of 'dusting out'.

A30 C

Ordered mixing occurs when micronised powders become adsorbed on to the surface of a larger 'carrier' particle. Segregation is therefore minimised while good flow properties are maintained. Segregation can occur if the carrier particles vary in size or if there are insufficient carrier particles, and the micronised particles may become dislodged if the mix is exposed to excessive vibration.

A31 D

Diffusion mixing has the potential to produce a random mix but generally results in a low speed of mixing.

A32   D

Tumbling mixers are good for free-flowing powders or granules but poor for cohesive or poorly flowing powders.

A33   C

Agitator mixers (e.g. ribbon mixer) are more difficult to clean than other mixer types. Ribbon mixers are good for mixing poorly flowing powders and are less likely to cause segregation than a tumbling mixer. 'Dead spots' are difficult to eliminate, and the shearing action of the blades may be insufficient to break up drug aggregates.

A34   A

The calculations assume that the particles are spherical, the particles have a limited particle size range, and a random mix is achieved. It is necessary to calculate the particle size of both the active ingredient and the excipients.

A35   B

A decrease in particle size will increase the number of particles in the scale of scrutiny and decrease the content variation. Decreasing the tablet weight will increase the proportion of active ingredient, thereby decreasing the content variation. Further increasing the mixing time will result in demixing. Further decreasing the particle size would present practical difficulties in handling the powder.

A36   B

Fine particles tend to be cohesive, exhibit poor flow properties and adhere to machine surfaces. Increasing the relative humidity to above 40% will decrease clumping due to static charges. Decreasing the particle size will increase the number of particles in the scale of scrutiny and decrease variation in content.

A37   E

Semisolids do not flow easily and rely on shear mixing, because of a lack of diffusion and convective mixing. 'Dead spots' are to be avoided since material will remain in these areas – hence a small clearance is required between the rotating elements and the mixing vessel in order to ensure sufficient shear. Planetary and sigma blade mixers are commonly used.

A38 C

Cohesion occurs between like surfaces (i.e. between component particles of a bulk solid), increases as particle size decreases, and is influenced by changes in the relative humidity and electrostatic forces.

A39 E

The bulk density is always less than the true density of a powder. An increase in interparticulate contacts causes an increase in cohesion and a decrease in porosity. Since powders normally flow under the influence of gravity, dense particles are generally less cohesive than less dense particles of the same size and shape. Void spaces between coarse particles may become filled with finer particles in a powder with a wide size range.

A40 E

The angle of repose is highest with cohesive particles and may have more than one angle of repose if it is very cohesive. Fine particles with high surface area to weight ratios are more cohesive. Dense particles are generally less cohesive than less dense particles of the same size and shape, since powders flow under the influence of gravity.

A41 A

Arches/bridges are formed more readily through the interlocking of non-isometric, highly textured particles. Electrostatic or cohesive forces can add to interparticulate attractions and promote closer particle packing, and arches/bridges may form in a closely packed powder.

A42 D

Porosity may be misleading when considering the packaging geometry, since arches/bridges may form in a closely packed powder; therefore, interparticulate pore size distribution must also be considered. Particle shape and texture and particle size distribution are more likely to influence packing geometries than particle size alone.

A43 A

Stresses acting on a stable arch are proportional to the width of the container. Powders with very low wall friction angles will empty freely. If a square-bottomed hopper is repeatedly refilled and partially emptied, the particles in the zone towards the base and sides will not be discharged. 'Rat-holes' are more likely in shallow-angled hoppers.

A44   A

Both the particle properties (including angle of repose and bulk density) and the design criteria of the hopper influence powder flow.

A45   D

Indirect methods for characterising powder flow include bulk density measurements, angle of repose, Carr's index and the Hausner ratio.

A46   D

Coarse particles are generally less cohesive, and a finer particle size is unlikely to improve flow. A very low RH may increase clumping due to static charges. HPMC is not a common glidant and has poor flow properties. Spray-drying can produce near-spherical particles that have improved flowability over irregularly shaped particles. Funnel flow may occur with hopper angles that are too shallow (i.e. with increased hopper angle).

A47   E

Vibration-assisted hoppers are useful in extreme cases where the hopper cannot be redesigned and the physical properties of the particles cannot be adjusted or the formulation altered.

A48   B

Increasing the length or speed of powder transfer pipes will increase electro-static charge and reduce powder flowability.

A49   C

A high angle of repose is seen with cohesive particles that have poor flow properties.

A50   B

In extreme cases, improving powder flow to improve weight uniformity may reduce content uniformity through increased segregation (demixing).

A51 B

Tablet/capsule machines fill by volume rather than weight. If different regions in the hopper contain granules of different sizes, and hence different bulk densities, a given volume in each region will contain a different weight of granules.

A52 E

It is important to control the particle size distribution of granules since although the individual components of a granule mix may not segregate, the granules themselves may segregate (i.e. demixing occurs). Granules are denser than the parent powder mix and occupy less volume per unit weight, making them more convenient for storage and shipping.

A53 D

Common granulating fluids include water, ethanol and isopropanol, either alone or in combination, and binders are frequently incorporated. Organic solvents are used for water-sensitive drugs as an alternative to dry granulation or where rapid drying times are required. Fluidised-bed granules have a greater porosity than those prepared by wet massing.

A54 C

If sufficient liquid is present in a powder to form a very thin, immobile layer, there will be an effective decrease in interparticulate distance, an increase in contact areas and an increase in bond strength. This contributes to the final granule strength in dry granulation.

A55 A

Continued kneading/mixing will increase the density of the wet mass, decrease the pore volume occupied by air, decrease interparticulate distance, increase granule tensile strength, and eventually produce the funicular or capillary state without further liquid addition.

A56 D

The capillary state is due to interfacial forces in mobile liquid films and is reached when particles are held by capillary suction at the liquid–air interface. The funicular state is an intermediate state between the pendular and capillary states.

A57 A

Fluidised-bed granulation is influenced by variables such as the fluidising air humidity and the temperature of the granulating fluid. Labour costs are minimised since all granulation processes are performed in one piece of equipment. Overmassing is encountered with high-speed mixer-granulators.

A58 E

Transfer losses occur when more than one piece of equipment is needed during the granulation process (e.g. shear granulators). High-speed mixer-granulators are capable of rapidly transforming usable granules into an unusable, overmassed system.

A59 C

Extrusion/spheronisation is a more labour-intensive process than other forms of granulation. It increases bulk density and flow properties, reduces the problems with dust usually encountered with finely divided powders, and has the ability to incorporate high levels of the active ingredient without producing large particles (i.e. minimum excipients are required).

A60 D

The most common application of extrusion is to produce spherical pellets for controlled release. The two major differences between the processes are that extrusion requires more granulating fluid and a uniform dispersion of the fluid is very important.

A61 B

Solute migration during drying of the wet spheres in the extrusion process will result in a stronger pellet with increased initial rate of dissolution and modified surfaces, which may reduce the adhesion of any added film coats.

A62 B

Once a solid reaches its equilibrium moisture content (i.e. only 'bound water' remains), extending the drying time will not change the moisture content unless the relative humidity of the ambient air is reduced. However, once the solid is exposed to humid ambient air, it will quickly regain moisture from the atmosphere. Since air transfers latent heat to a wet solid as it cools, the temperature of the air can fall to dew point with excessive cooling. Tablet granules have superior compaction properties with a small amount (1–2%) of residual moisture.

A63   A

Freeze drying results in a final dry product that is light and porous, since it is a network of solid occupying the same volume as the original solution. Bulk density is increased with spray drying.

A64   C

Drying occurs at low temperatures and under vacuum, and therefore hydrolysis and oxidation are minimised. The final product is hygroscopic and therefore packaging requires special consideration. Since there is no concentration of the solution before drying, salts do not concentrate in the wet state. The porous nature of the product improves the solubility of the freeze-dried product.

A65   E

Fluidised-bed drying is a convection method of drying requiring a short drying time. Aggregation of particles is unlikely due to the turbulent air flow, but the turbulence may produce dust and generate static electricity charges, which are particularly dangerous if organic materials or organic solvents are used.

A66   C

Microwave drying utilises radiation as a heat transmission method and is found to be efficient in the wavelength range of 10 mm to 1 m. A slight vacuum is used in industry since the airflow through the chamber facilitates the continuous removal of evaporated solvent. The batch size of commercial production microwave dryers is smaller than batch sizes available for fluidised-bed drying.

A67   B

If intragranular migration is likely, vacuum or microwave dryers may be more suitable than fluidised-bed drying. Tray drying (static convective drying) is more likely to cause intergranular migration. The smallest granules that will flow easily are preferable if mottling is a problem. Aluminium lakes reduce mottling compared with water-soluble dyes. Increasing the viscosity of the granulating fluid will decrease solute migration.

A68   C

Granulation of a powder is the reverse of size reduction of a bigger particle. Because of the closer binding of powder particles, granules have a higher density. A reduction in surface area reduces the moisture absorption.

A69   A

For producing spheres, the powder mass is kneaded with a binder and extruded from an extruder; exudates are spheronised; and the spheres produced are dried and finally sieved to get different size portions.

A70   E

The granulation process helps improve flow, prevents segregation by forming compacts, and reduces dusting and moisture absorption.

A71   C

Coating improves aesthetics, stability and compliance, while masking unpleasant colour, odour and taste.

A72   D

The sugar-coating process usually takes 2–5 days because of the number of steps involved. A shorter time requirement is a major advantage offered by film coating.

A73   A

In sugar coating, the tablet surface is sealed to prevent permeation of moisture to the core, subcoated to round the edges, grossed with heavy syrup, colour coated and polishing after drying.

A74   D

Potassium chloride is not a polymer and will not provide the sealing effect required in sugar coating.

A75   E

Sugar coating is a complex process with a large number of variables to be optimised. This is critical in order to achieve reproducible colour and function.

A76 C

A pentagonal shape will be difficult to convert to a roundish tablet and will require heavy subcoating before the final coating. A light blue colour can be another less desirable property, but it is easier to fix.

A77 C

Film coating does not require sealing similar to sugar coating. Film coating is a preferred option for non-spherical tablets.

A78 D

All of the listed excipients categories are usually used in film coating.

A79 D

Enteric coating can help in avoiding the acidic pH of the stomach, ensuring stability of the drug.

A80 C

Polyvinyl alcohol does not provide enteric coating and will not protect from an acidic pH.

A81 D

Solvents are very carefully monitored by regulatory authorities and pharmaceutical companies.

A82 B

All types of capsule can be coated by modifying the coating composition and process.

A83 C

10% w/w refers to 10 g of the drug in a total weight of solution of 100 g, leading to 5 g in 50 mL.

A84 C

25 ng/mL is equivalent to 2.5 µg/100 mL.

$2.5 \text{ µg} = 2.5 \times 10^{-3} \text{ mg} = 2.5 \times 10^{-6} \text{ g per } 100 \text{ mL}$

A85 B

This is an example of a dilution experiment. As the concentration of solution provided is only 90%, the volume required can be worked out by calculating $(100 \times 70)/90 = 77.8$.

A86 C

One part per million refers to 1 g per 1 million mL, i.e. 1 part in 10^6 parts.

A87 A

% w/w refers to the weight per 100 g. Therefore, as the quantity of drug in 25 g is 1 g, the % w/w is 4%.

A88 B

One mole contains 496 g and 1 M solution contains 496 g in 1000 mL. For 0.2 M solution, you will need $496/5 = 99.2$ g for 1000 mL and 9.9 g for 100 mL.

A89 A

10% w/v contains 10 g in 100 mL. The calculation is completed by scaling up the quantities and volumes, i.e. 200 kg for 2000 L.

A90 D

The calculation is performed by initially determining the quantity of drug required in 100 L to produce a 10% w/v solution. The actual amount is subtracted from this nominal amount to give the quantity of drug required. This is then dissolved in 10 L. The concentration required to adjust the shortfall is 0.85%. Adding 10% of the additional 10 L, the final percentage required is 18.5%.

A91 D

This calculation is the reverse of Q90. Having worked out the overage quantity, the volume of diluent is then calculated to produce the required concentration. Remember that by adding diluents, the overall volume of the batch has increased.

A92   C

There are two stages to this calculation. Initially work out the quantity of drug required to produce a 200 kg batch assuming 100% purity. Then divide this by 0.92 to correct for purity, i.e. $(200 \times 5)/0.92 = 10.87$.

A93   C

This question is calculated in a similar fashion to the previous question. Remember that the question asks for a 10% w/v solution of oxytetracycline (it requires 496 g of the hydrochloride to produce 460 g of oxytetracycline). Therefore, initially calculate the quantity of oxytetracycline required and then convert for purity.

A94   C

A 1 in 5000 solution contains 1 g in 5000 mL. The stock solution is diluted ten times and therefore was originally 1 in 500, i.e. 1 g in 500 mL.

A95   A

A 1 in 5000 solution contains 1 g in 5000 mL, i.e. 200 mL will contain 0.04 g.

A96   B

Scale up the calculation from 1 h to 24 h and from 1 kg to 50 kg, i.e. $10 \times 50 \times 24 = 12\,000 = \mu g$, or 12 mg.

A97   D

This is a standard dilution equation, where the dose is to be diluted four times and the quantity required is $100/4 = 25$ g.

A98   C

A 1% w/v solution refers to one that contains 1 g dissolved in a final volume of 100 mL.

A99   A

5% w/w of drug refers to 5 kg in 100 kg, i.e. 25 kg for 500 kg.

A100   B

0.3% w/w of preservative refers to 0.3 kg in 100 kg, i.e. 1.5 kg for 500 kg.

Test 4

Dosage form design

Therese Kairuz, Sanjay Garg, Paul Chi-Lui Ho

and Roop K Khar

Liquid dosage forms and clarification	Tablets, compaction
Emulsions and suspensions	Capsules
Semisolids	

Introduction

Conventional dosage forms, including liquids, semisolids and solids, are the backbone of pharmaceutical industry and practice. These form the core of the pharmaceutics curriculum around the world, at undergraduate and postgraduate levels, although there are subtle differences in the depth and breadth of topics covered in different countries. In countries with strong pharmaceutical industries, such as India and China, emphasis is on related equipments and processing. In countries with a limited industrial basis, the focus is more on clinical aspects, extemporaneous compounding and dosage regimens. Irrespective of the scope, a good understanding of these dosage units is critical for every pharmacist.

Drug solubility and solubilisation dictate choice of dosage form, bioavailability and therapeutic activity. Solutions are the simplest dosage form and are preferred when possible. Emulsions and suspensions offer good options for insoluble drugs. Excipients such as surfactants, viscosity-enhancing agents, preservatives, colours and flavours help in achieving desired product characteristics. The choice of surfactant is governed by the type of emulsion and the proportion of lipid and aqueous phases and other excipients present in the formulation. Hydrophilic lipophilic balance (HLB) provides a good guide to the selection of appropriate surfactant for a given system.

Semisolids such as gels, creams and ointments offer topical alternatives for drug delivery to local and systemic sites. Creams are essentially biphasic systems in the form of water-in-oil (w/o) or oil-in-water (o/w) emulsions, most suited for water-insoluble drugs. Gels are more commonly used for water-soluble drugs, while ointments are the preferred dosage forms for hydrophobic drugs.

In pharmaceuticals, the oral route is the primary route of drug delivery, and tablets are the most convenient unit dose option. Tablets are easy to develop and scale up to commercial quantities, store and administer. As tablets are solid dosage forms with a low moisture content, drug stability is better, especially for water-sensitive drugs. The range of tablet design offers good flexibility at development stages. Tabletting technology has evolved significantly over the years, from simple single-punch compression machines to fully automated high-speed systems operating under a good manufacturing practices (GMP) environment. Hard and soft gelatin capsules are also very commonly used, offering similar flexibility and design options. Although the capsule shells are still mostly based on gelatin, some options such as hydroxypropyl methylcellulose-based vegetarian capsules are becoming increasingly popular. The flexibility of delivering drugs in the form of conventional and controlled-release tablets and capsules has widened the scope of practice of pharmacy.

A range of excipients, including vehicles, binders, disintegrants, glidants, lubricants, sweeteners, preservatives, colours and flavours, are used to design solid dosage forms. Advances in excipients such as spray-dried materials and super-disintegrants have simplified the manufacturing process. From the practice point of view, an understanding of different dosage forms and their composition, stability issues and flexibility of extemporaneous compounding is highly useful.

Questions

Q1   An aqueous solution can be readily formulated with:

A ❑ magnesium chloride
B ❑ sodium stearyl fumarate
C ❑ magnesium stearate
D ❑ methylhydroxybenzoate
E ❑ methylcellulose

Q2   Water is widely used as a pharmaceutical solvent because of its:

A ❑ low dielectric constant
B ❑ high buffering capacity
C ❑ selective solvency potential

D ❑ physiological compatibility

E ❑ organic matter extraction capability

Q3   Solubility is determined by the:

A ❑ extent of dissolution that occurs

B ❑ rate at which dissolution occurs

C ❑ equilibrium that occurs between solutes and solvent

D ❑ saturation rate of the solution

E ❑ miscibility of the solute and solvent

Q4   Which of the following is true regarding co-solvents?

A ❑ they are immiscible with the continuous phase

B ❑ they are required to be less than 1% of the total continuous phase volume

C ❑ they are used to dissolve non-polar compounds

D ❑ they have dielectric constants that are equal to that of the solvent

E ❑ they are used to improve the solubility of strong electrolytes

Q5   Which of the following is true regarding Phytomenadione?

A ❑ it is a water-soluble vitamin required for clotting

B ❑ it is an analogue of warfarin

C ❑ it has surfactant properties that facilitate its inclusion in aqueous syrups

D ❑ it has an HLB value of 2

E ❑ it needs to be solubilised in an aqueous vitamin preparation

Q6   Which of the following is true regarding glycerol?

A ❑ it is not recommended as a co-solvent for use in children due to potential toxicity

B ❑ it is not used as a co-solvent in aqueous solutions, only in non-polar solutions

C ❑ it is not used together with ethanol as a co-solvent because of immiscibility

D ❑ it has a sweet taste

E ❑ it has no preservative function in aqueous mixtures

Q7   Select the correct statement from the following:

A ❑ absorption from an aqueous solution avoids first-pass metabolism

B ❑ absorption from an aqueous solution is slower than from a suspension

C ❏ absorption from a solution occurs in the intestine
D ❏ dilution may cause precipitation with some drugs
E ❏ dilution increases pourability

Q8 Which of the following will result in reduced solubility?

A ❏ increased number of carbons in a chain
B ❏ increased branching of chains
C ❏ increased hydrogen bonding
D ❏ increased ratio of polar/non-polar groups
E ❏ increased solvation

Q9 Which of the following devices should not be used to measure a paediatric dose?

A ❏ oral syringe
B ❏ calibrated spoon
C ❏ graduated dropper
D ❏ calibrated cup
E ❏ teaspoon

Q10 Which of the following is not an advantage of aqueous solutions?

A ❏ homogeneous dispersion of the active ingredient
B ❏ reduced need for preservatives
C ❏ easy identification of products through the use of suitable colorants
D ❏ accurate dosing of water-soluble drugs
E ❏ reduced difficulties associated with dysphagia

Q11 Which of the following products is not available as a solution?

A ❏ gargle
B ❏ nasal spray
C ❏ reconstituted antibiotic for oral use
D ❏ enema
E ❏ mouthwash

Q12 Which of the following liquids is not used as a vehicle for paediatric oral solutions?

A ❏ ethanol
B ❏ propylene glycol
C ❏ glycerin

D ❏ syrup
E ❏ commercially available bases with sweetening and
suspending properties

Q13 Which of the following approaches would not be suitable for improving drug solubility with an aim of increasing the dissolution rate?

A ❏ complexation
B ❏ micellar solubilisation
C ❏ salt form
D ❏ hydrated form
E ❏ co-solvents

Q14 Which of the following statements about the Noyes–Whitney
dissolution mechanism is incorrect?

A ❏ the dissolution rate is directly proportional to the surface
area of the drug particle
B ❏ the dissolution rate is not equal to the intrinsic dissolution
rate
C ❏ the drug concentration in the diffusion layer is higher than in
the unstirred layer
D ❏ the dissolution rate is dependent on the drug concentration
in the bulk solution
E ❏ the drug dissolution rate can be measured using Wood's
apparatus

Q15 Which of the following statements is (are) correct? An aqueous oral
solution may include:

1 ❏ solubilised hydrophobic substances
2 ❏ oil in a 60 : 40 oil to water ratio
3 ❏ micelles
4 ❏ insoluble solutes
A ❏ 1 and 3
B ❏ 1 and 2
C ❏ 1, 2 and 3
D ❏ 3
E ❏ 3 and 4

Q16 Which of the following statements is (are) correct? The dissolution
rate from a solution depends on the:

1 ❏ wettability of the drug

     2   ❑ surface area of the particles
     3   ❑ hydrophobic lipophilic balance
     A ❑ 1 and 2
     B ❑ 2 and 3
     C ❑ 1
     D ❑ 2
     E ❑ 3

Q17   Which of the following statements is (are) correct? A solution:

     1   ❑ is a mixture of two or more components
     2   ❑ is homogeneous down to molecular level
     3   ❑ contains a disperse phase of solutes
     4   ❑ has the largest phase as the solvent
     A ❑ 1
     B ❑ 2
     C ❑ 1, 2 and 3
     D ❑ 3 and 4
     E ❑ 1 and 2

Q18   Which of the following statements is (are) true?

     1   ❑ potable water is suitable for oral pharmaceutical solutions
     2   ❑ purified water is prepared by reverse osmosis
     3   ❑ purified water is prepared by distillation
     4   ❑ pyrogen-free water is prepared by heating with a bactericide
     A ❑ 1
     B ❑ 2
     C ❑ 3
     D ❑ 1, 2 and 3
     E ❑ 3 and 4

Q19   Sterility testing for detecting the presence of viable microbes in pharmacopoeial preparations is based on:

     A ❑ distillation
     B ❑ evaporation
     C ❑ drying
     D ❑ centrifugation
     E ❑ filtration

Q20 Of the following filtration assemblies, the one that can be used for continuous operation is:

A ❏ rotary drum filter
B ❏ filter leaf
C ❏ Sweetland filter
D ❏ filter press
E ❏ edge filter

Q21 The porosity of the sintered glass filters depends on the:

A ❏ quality of the glass granules
B ❏ size of the glass granules
C ❏ quantity of the glass granules
D ❏ melting point of the glass granules
E ❏ all of the above

Q22 Separation of liquid–liquid mixtures can be carried out by using:

A ❏ Nutsche filters
B ❏ filter leaves
C ❏ filtering centrifuges
D ❏ centrifuge sedimentors
E ❏ membrane filters

Q23 An example of a filter aid is:

A ❏ glass
B ❏ cotton
C ❏ talc
D ❏ filter paper
E ❏ porcelain

Q24 In Darcy's equation (rate of filtration = $kA\Delta P/L\eta$), Darcy's constant or permeability coefficient (k) depends upon:

A ❏ area of filter media, A
B ❏ pressure difference across the filter medium, ΔP
C ❏ thickness of the filter medium, L
D ❏ viscosity, η
E ❏ nature of the filter medium

Q25 The rate of filtration increases with an increase in:

1 ❏ pressure difference

2 ❑ area of filter
3 ❑ viscosity
A ❑ 1, 2 and 3
B ❑ 1 and 2
C ❑ 2 and 3
D ❑ 1 and 3
E ❑ 2

Q26   Which of the following statements is (are) correct? Filters are validated by the:

1 ❑ bubble point method
2 ❑ cloud point method
3 ❑ flash point method
A ❑ 1, 2 and 3
B ❑ 1 and 2
C ❑ 2 and 3
D ❑ 1
E ❑ 2

Questions 27 and 28 involve the following case:

In a product development laboratory, a scientist found colloidal solids as impurities in a liquid formulation. The concentration of colloidal solids was less than 1.0% w/v.

Q27   The unit operation that can be used for the removal of solids is:

A   ❑ clarification
B   ❑ filtration
C   ❑ ultrafiltration
D   ❑ surface filtration
E   ❑ drying

Q28   In the selected unit operation, flow of fluid through interstices of particles is expressed by:

A   ❑ Poiseuille's law
B   ❑ Kermicer equation
C   ❑ Henderson equation
D   ❑ Grenier value
E   ❑ Stokes' equation

Q29 Which of the following factors does not influence the bioavailability of drugs from aqueous suspension dosage forms?

A ❑ particle size
B ❑ crystal form of the drug
C ❑ inclusion of surfactants as wetting and deflocculating agents
D ❑ solubility of the drug
E ❑ none of the above

Q30 Which of the following surface active agents is (are) used as wetting agents for oral administration?

1 ❑ polysorbates
2 ❑ sorbitan esters
3 ❑ sodium lauryl sulfate
A ❑ 1
B ❑ 2
C ❑ 3
D ❑ 1 and 2
E ❑ 2 and 3

Q31 Which of the following surface active agents is (are) used as wetting agents for parenteral administration?

1 ❑ polysorbates
2 ❑ sorbitan esters
3 ❑ sodium lauryl sulfate
A ❑ 1
B ❑ 2
C ❑ 3
D ❑ 1 and 2
E ❑ 2 and 3

Q32 When a suspension is formulated, it has to be:

A ❑ deflocculated (i.e. the dispersed particles remain as discrete units) and with controlled viscosity
B ❑ deflocculated and without controlled viscosity
C ❑ flocculated and without controlled viscosity
D ❑ partially flocculated and with controlled viscosity
E ❑ none of the above

Q33 Which of the following statements about alginates as viscosity modi-fiers is (are) correct?

1 ❑ alginate mucilage must be heated above 60 °C during preparation to induce its viscosity
2 ❑ alginates are most viscous around 24 h after preparation
3 ❑ at low pH, alginates will precipitate
   A ❑ 1
   B ❑ 2
   C ❑ 3
   D ❑ 1 and 2
   E ❑ 2 and 3

Q34 Surfactants suitable for use as wetting agents in the formulation of suspensions would possess an HLB value in the range of:

A ❑ 2–4
B ❑ 5–6
C ❑ 7–9
D ❑ 10–13
E ❑ 14–16

Q35 Which of the following statements regarding methylcellulose as a viscosity modifier is incorrect?

A ❑ it is non-ionic
B ❑ it is stable over the pH range 5–10
C ❑ it is compatible with many ionic additives
D ❑ on heating, it will become progressively dehydrated and gel at 50 °C
E ❑ none of the above

Q36 Which of the following statements about sodium carboxymethyl-cellulose as a viscosity modifier is correct?

A ❑ it is non-ionic
B ❑ it is stable over the pH range 5–10
C ❑ it is compatible with many ionic additives
D ❑ on heating, it will become progressively dehydrated and gel at 50 °C
E ❑ none of the above

Q37 When a suspension is applied externally, the texture will feel gritty to patients if the particle size (diameter, in μm) is greater than:

A ❏ 1
B ❏ 5
C ❏ 10
D ❏ 20
E ❏ 100

Q38 Which of the following viscosity modifiers can be used internally?

1 ❏ acacia
2 ❏ tragacanth
3 ❏ bentonite
   A ❏ 1
   B ❏ 2
   C ❏ 3
   D ❏ 1 and 2
   E ❏ 1, 2 and 3

Q39 Which of the following is (are) used in the formulation of an oral emulsion?

1 ❏ liquid paraffin
2 ❏ castor oil
3 ❏ benzyl benzoate
   A ❏ 1
   B ❏ 2
   C ❏ 3
   D ❏ 1 and 2
   E ❏ 1, 2 and 3

Q40 Which of the following oils is (are) used for intravenous administration in the formulation of an emulsion?

1 ❏ cod liver
2 ❏ cottonseed
3 ❏ safflower
   A ❏ 1 and 2
   B ❏ 2 and 3
   C ❏ 1 and 3
   D ❏ 1, 2 and 3
   E ❏ none of the above

Q41 Which of the following statements about emulgents is (are) correct?

1 ❏ non-ionic emulgents are less irritant than anionic emulgents
2 ❏ cationic emulgents can be administered orally at low concentration
3 ❏ cationic emulgents can be antiseptics
   A ❏ 1
   B ❏ 2
   C ❏ 3
   D ❏ 1 and 3
   E ❏ 2 and 3

Q42 Which of the following approaches does not improve the physical stability of emulsions?

A ❏ production of an emulsion of small droplet size
B ❏ increase in the viscosity of the continuous phase
C ❏ increase in the density difference between the two phases
D ❏ control of the disperse phase concentration
E ❏ none of the above

Q43 Many of the oils and fats used in emulsions are susceptible to:

A ❏ hydrolysis
B ❏ oxidation
C ❏ reduction
D ❏ pyrolysis
E ❏ thermal degradation

Q44 Which of the following statements about emulsions is incorrect?

A ❏ emulsions with a droplet size of less than 1 μm could be transparent
B ❏ emulsions for intravenous administration must be of the oil-in-water type
C ❏ water-in-oil emulsions are more suitable for hydrating skin
D ❏ water-in-oil emulsions are more efficient cleansers
E ❏ none of the above

Q45 What is the HLB value of a combination of surfactants composed of 2 g Span 80 (HLB = 4.3) and 3 g Tween 80 (HLB = 15)?

A ❏ 53.6
B ❏ 19.3

C ❏ 18.9
D ❏ 10.7
E ❏ 11.5

Q46 What would be the amount (in g) of Span 80 in 5 g of a surfactant system composed of a mixture of Span 80 (HLB = 4.3) and Tween 80 (HLB = 15) to produce a total HLB value of 13?

A ❏ 0.54
B ❏ 0.93
C ❏ 1.23
D ❏ 1.52
E ❏ 3.26

Q47 Which of the following is not used as a preservative for emulsions?

A ❏ sorbic acid
B ❏ chlorocresol
C ❏ bronopol
D ❏ parahydroxybenzoic acid esters
E ❏ chloral hydrate

Q48 For the preparation of an acacia emulsion, the proportion of the amounts of oil, water and acacia in the primary emulsion should be:

A ❏ 4:2:1
B ❏ 4:4:2
C ❏ 4:8:4
D ❏ 2:4:1
E ❏ 4:4:1

Q49 Select the correct option with reference to excipients in creams:

1 ❏ creams without a preservative should be given a short shelf life
2 ❏ creams do not require a preservative because of their high oil content
3 ❏ due to partitioning, additional preservatives may be required in creams
4 ❏ due to partitioning, higher concentrations of preservatives may be required in creams

A ❏ 1
B ❏ 1 and 2

C ❑ 1, 2 and 3
D ❑ 3 and 4
E ❑ 1, 3 and 4

Q50 Select the option that describes the characteristics of a water-miscible semisolid base?

1   ❑ dehydration of the skin
2   ❑ inactivation of some preservatives
3   ❑ poor release of drug
4   ❑ difficult to remove from the skin
A ❑ 1 and 2
B ❑ 2 and 3
C ❑ 1 and 4
D ❑ 1, 2 and 3
E ❑ 3 and 4

Q51 Select the option that correctly describes the emulsification of creams:

1   ❑ wool fat is used to form water-in-oil emulsions
2   ❑ wool alcohol reduces the 'water-holding' capacity of greasy bases
3   ❑ cetostearyl alcohol forms water-in-oil emulsions
4   ❑ emulsifying waxes form water-in-oil emulsions when water is added
A ❑ 1
B ❑ 1 and 2
C ❑ 1, 2 and 3
D ❑ 3 and 4
E ❑ 4

Q52 Select the option that best describes the properties of ointments:

1   ❑ medicated ointments can be used to treat conditions of the eye, nose, vagina and anus
2   ❑ treatment with ointments is for local administration
3   ❑ ointments can be used as protectants
4   ❑ anhydrous ointments are the semisolid of choice for oozing wounds
A ❑ 1 and 2
B ❑ 1 and 3
C ❑ 1, 2 and 3
D ❑ 3 and 4
E ❑ 4

Q53 Which of the following is incorrect with reference to creams?

A ❑ cetomacrogol can be added as a surfactant

B ❑ sodium lauryl sulfate is often included as an anionic surfactant

C ❑ cetostearyl alcohol normally functions as a surfactant

D ❑ soft paraffin often provides the oil phase

E ❑ cetomacrogol is often included as a water-in-oil emulsifier

Q54 Which of the following is incorrect with reference to creams?

A ❑ creams consist of at least two phases

B ❑ cetrimide is often included as an anionic surfactant

C ❑ mixed emulsifier systems often contain cetostearyl alcohol

D ❑ creams are semisolid emulsions for external application

E ❑ creams are thermodynamically unstable

Q55 Which of the following is incorrect with reference to creams?

A ❑ creams may contain preservatives, emulsifiers, antioxidants and volatile propellants

B ❑ cetomacrogol cream is the diluent of choice for betamethasone, a cationic drug

C ❑ aqueous cream is anionic and cetrimide cream is cationic

D ❑ creams may contain a co-solvent to prevent drug precipitation

E ❑ phase inversion can occur when a semisolid emulsion is applied to the skin

Q56 Which of the following is incorrect with reference to creams?

A ❑ drugs may be trapped in micelles within the semisolid emulsion

B ❑ a standard amount of preservative is used for creams

C ❑ oil-in-water semisolid emulsions are used for their 'washability' off the skin

D ❑ oil-in-water semisolid emulsions are used for their 'spreadability' on the skin

E ❑ oil-in-water creams are used for their 'vanishing' properties on the skin

Q57 Which of the following products would not spread readily on the skin?

A ❑ lotion

B ❑ cream

C ❏ ointment
D ❏ paste
E ❏ foam

Q58 Which of the following cannot be used to determine the shelf life of external preparations?

A ❏ accelerated stability test at room temperature over a prolonged time
B ❏ accelerated stability test at elevated temperatures
C ❏ adjustments to account for loss of volatile ingredients
D ❏ Arrhenius equation
E ❏ Fick's law

Q59 Which of the following is not true with reference to preservative use in a dermatological formulation?

A ❏ the preservative should be compatible with all of the ingredients
B ❏ the preservative should be soluble in the oil phase
C ❏ the preservative should be stable to heat
D ❏ the preservative should be non-irritant and non-toxic to human tissue
E ❏ the preservative should be with a known octanol/water partition coefficient

Q60 Which of the following is not a potential source of microbial contamination of topical products?

A ❏ *Candida albicans*
B ❏ raw materials
C ❏ processing equipment
D ❏ distilled water
E ❏ plant and environment hygiene

Q61 Which of the following will not affect the stability of an acidic active ingredient?

A ❏ use of a pH electrode
B ❏ change in pH from 6.6 to 3.6
C ❏ solubilisation in 1.0 M sodium hydroxide solution
D ❏ increase of 10 °C in the storage temperature
E ❏ decrease of 10 °C in the storage temperature

Q62   Ointments:

A   ❑ are anhydrous preparations for external use
B   ❑ require a preservative at a concentration higher than creams
C   ❑ may contain dissolved or dispersed medicaments
D   ❑ may be made with water-soluble bases
E   ❑ contain lanolin (wool fat) as an ingredient in many cases

Q63   Which of the following is not true about skin?

A   ❑ it is the largest organ in the body
B   ❑ it contains blood vessels, nerve fibres, sweat glands and hair follicles in the dermis
C   ❑ it has an outermost region, the epidermis, which is a single layer of cells
D   ❑ the stratum corneum consists of dead, keratinised cells
E   ❑ hair follicles and sweat glands originate in the dermis and extend to the epidermis

Q64   Which of the following pair of ointment ingredients refer to the same thing?

A   ❑ light liquid paraffin and light mineral oil
B   ❑ liquid paraffin and mineral oil
C   ❑ soft paraffin and petrolatum
D   ❑ soft paraffin and petroleum
E   ❑ paraffin and hard paraffin

Q65   Which of the following products is an example of a topical preparation?

A   ❑ turpentine liniment
B   ❑ benzyl benzoate application
C   ❑ compound hydroxybenzoate solution
D   ❑ compound sulfur lotion
E   ❑ sterile sodium chloride solution

Q66   The topical route is most commonly used for:

A   ❑ local effects only
B   ❑ systemic effects for drugs that require metabolism
C   ❑ local and systemic effects
D   ❑ active ingredients that show zero-order kinetics
E   ❑ active ingredients that are required for an immediate onset of action

Q67 Which of the following is correct regarding the dilution of creams? The dilution of creams:

A ❑ is recommended for use in children
B ❑ is required for cortisone preparations
C ❑ should occur with aqueous cream
D ❑ affects the bioavailability of the medicament
E ❑ doubles the concentration of the preservative

Q68 Which of the following statements is incorrect?

A ❑ creams are emulsified systems
B ❑ ointments are hydrophilic
C ❑ pastes contain a high concentration of solids
D ❑ pastes exhibit non-Newtonian flow behaviour
E ❑ lotions require a preservative

Q69 During the preparation of creams, the aqueous and non-aqueous phases are separately heated and mixed with stirring. What is the typical range of temperature (in °C)?

A ❑ 30–40
B ❑ 40–50
C ❑ 55–70
D ❑ 80–90
E ❑ 120–150

Q70 Which of the following is true regarding topical and transdermal formulations?

A ❑ topical formulations provide controlled release, while transdermal preparations provide immediate release of the drug
B ❑ transdermal formulations are more effective as sunscreen compared with topical formulations
C ❑ topical formulation are for localised action, while transdermal preparations are designed for systemic action
D ❑ there is no difference between topical and transdermal formulations
E ❑ transdermal preparations are usually very simple, while topical preparations tend be complex in composition

Q71 Which of the following statements is not valid for gels?

A ❑ gels are usually transparent and water based

B ❑ in the preparation of gels, aqueous and non-aqueous phases should be heated to the same temperature

C ❑ syneresis is a form of instability in gels

D ❑ polymer–solvent interaction provides the required three-dimensional structure in gels

E ❑ a gel cannot be formed below a critical concentration of polymer

Q72 Which of the following non-ionic surfactants will be the best choice for preparation of a water-removable ointment base?

A ❑ sorbitan trioleate (Span 85), HLB 1.8

B ❑ glyceryl monostearate, HLB 3.8

C ❑ sorbitan monostearate (Span 60), HLB 4.7

D ❑ polyoxyethylene sorbitan trioleate (Tween 85), HLB 11.0

E ❑ a new surfactant, HLB 7.0

Q73 There are four types of ointment base: hydrocarbon, absorption, water-removable and water-soluble. Which of these provides good occlusion?

A ❑ hydrocarbon

B ❑ water-removable

C ❑ absorption

D ❑ water-soluble

E ❑ hydrocarbon and absorption

Q74 Which of the following is not true with reference to the popularity of tablets?

A ❑ being a unit dosage form, tablets offer accuracy of dose

B ❑ the oral route is the most convenient and safe way of administering drugs

C ❑ tablets are easy to produce on a bulk scale

D ❑ tablets are usually costlier compared with oral liquid formulations

E ❑ tablets offer better stability for most drugs compared with liquid dosage forms

Q75 The process of tablet formation involves three stages. Which of the following is the correct sequence?

A ❑ die filling > compression > ejection

B ❑ compression > die filling > ejection

C ❏ die filling > ejection > compression
D ❏ ejection > die filling > compression
E ❏ compression > ejection > die filling

Q76 A number of powder properties need to be optimised and controlled in the tabletting process. Which of the following options best describes these properties?

1 ❏ homogeneous mixing of powder with minimal segregation
2 ❏ compression characteristics
3 ❏ flow behaviour
4 ❏ adhesion and friction
    A ❏ 1 and 2
    B ❏ 2 and 3
    C ❏ 3 and 4
    D ❏ 1, 2 and 3
    E ❏ 1, 2, 3 and 4

Q77 Which of the following is not the correct pair of equipment and unit operation?

A ❏ high shear mixer: mixing of powder
B ❏ high shear mixer: agglomeration
C ❏ fluidised-bed dryer: size reduction
D ❏ hammer mill: size reduction
E ❏ tablet press: tabletting

Q78 Which of the following is true regarding direct compression?

A ❏ the process of direct compression is as complex as wet granulation
B ❏ direct compression is cheaper than wet granulation and compression
C ❏ there is no difference in stability between tablets produced by any method
D ❏ any material can be directly compressed with modern tabletting equipment
E ❏ direct compression cannot be used for potent drugs

Q79 Which of the following tablet excipients is not used as a filler?

A ❏ lactose
B ❏ calcium carbonate
C ❏ dicalcium phosphate

D ❑ sorbitol
E ❑ polyvinyl pyrrolidone

Q80 Which of the following is the most commonly used lubricant in tablet manufacturing?

A ❑ polyethylene glycol
B ❑ liquid paraffin
C ❑ magnesium stearate
D ❑ stearic acid
E ❑ sodium lauryl sulfate

Q81 Which of the following can be used as a dry binder?

A ❑ gelatin
B ❑ sucrose
C ❑ starch
D ❑ polyvinyl pyrrolidone
E ❑ sodium starch glycolate

Q82 Which of the following mechanisms can facilitate tablet disintegration?

1 ❑ swelling of disintegrants by sorption of water, leading to rupture of the tablet
2 ❑ promotion of penetration of liquid inside the tablets, such as that caused by surface active agents
3 ❑ evolution of gas such as carbon dioxide, in the presence of water

A ❑ 1
B ❑ 2
C ❑ 3
D ❑ 1 and 2
E ❑ 1, 2 and 3

Q83 Which of the following is incorrect with reference to lubrication during tabletting?

A ❑ glidants improve flowability during the tabletting process
B ❑ colloidal silica is the most commonly used glidant
C ❑ lubricants reduce the friction between tablet and die, helping with compression and ejection
D ❑ lubricants also help with tablet disintegration and dissolution
E ❑ magnesium stearate is an example of a commonly used lubricant

Q84 Which of the following types of tablet can be used for the sustained release of drugs?

A ❑ disintegrating
B ❑ chewable
C ❑ effervescent
D ❑ buccal
E ❑ sublingual

Q85 Which of the following is not true with reference to lozenges?

A ❑ compressed lozenges are designed to dissolve slowly and release the drug in the saliva
B ❑ lozenges are often coloured and flavoured
C ❑ similar to other tablets, lozenges contain gelatin as the disintegrating agent
D ❑ glucose and sorbitol are commonly used as fillers in lozenges
E ❑ lozenges are generally produced by compression at high pressure

Questions 86 and 87 involve the following case:

A formulation scientist is asked to develop a cost-effective vaginal dosage form for tropical countries. The drug is highly water-soluble at pH 4.0 and is designed for local vaginal infection. The immediate release of defined amount of drug and prolonged retention in the vagina is desirable for optimal biological effect.

Q86 Which of the following is the best formulation option?

A ❑ cocoa butter suppository
B ❑ vaginal tablet
C ❑ vaginal capsule
D ❑ foam
E ❑ solution for vaginal application

Q87 If a solid dosage form is selected as an option, which of the following properties should be included in the design?

A ❑ rapid disintegration
B ❑ stability at temperatures above body temperature
C ❑ bioadhesion

D ❑ non-irritant formulation

E ❑ all of the above

Q88 Which of the following can be used as a binder in aqueous and non-aqueous systems?

A ❑ polymethacrylates

B ❑ acacia

C ❑ starch

D ❑ sucrose

E ❑ alginic acid

Q89 Which of the following is (are) used as drug-elease mechanisms in prolonged-release tablets?

1 ❑ diffusion control

2 ❑ dissolution control

3 ❑ erosion control

A ❑ 1

B ❑ 2

C ❑ 3

D ❑ 1, 2 and 3

E ❑ 1 and 2

Q90 Which of the following is not the correct pair of test parameter and method used for assessment?

A ❑ uniformity of content: assay of active ingredient

B ❑ mechanical strength: hardness test

C ❑ release profile: disintegration test

D ❑ microbial load: sterility testing

E ❑ mottling: surface appearance

Q91 Which of the following best describes the reasons for carrying out dissolution studies?

1 ❑ to estimate the likely in vivo behaviour, based on in vitro data

2 ❑ to act as a quality-control tool

3 ❑ to develop generic formulations for specific brands

A ❑ 1

B ❑ 2

C ❑ 3

D ❑ 1, 2 and 3

E ❑ 1 and 3

Q92   Which of the following is the correct sequence of two-piece hard gelatin capsule shell manufacturing?

A   ❑ dipping > spinning > drying > stripping > cutting > joining
B   ❑ dipping > stripping > spinning > drying > cutting > joining
C   ❑ dipping > spinning > stripping > drying > cutting > joining
D   ❑ dipping > spinning > stripping > cutting > joining > drying
E   ❑ dipping > spinning > drying > stripping > joining > cutting

Q93   Which of the following capsule sizes will provide the largest body fill volume?

A   ❑ 0
B   ❑ 1
C   ❑ 2
D   ❑ 3
E   ❑ 4

Q94   Which of the following is normally not filled in hard gelatin capsules?

A   ❑ powders
B   ❑ granules
C   ❑ tablets
D   ❑ aqueous solutions
E   ❑ pastes

Q95   Which of the following is not true with reference to capsule filling and formulation?

A   ❑ similar to tablets, capsules also contain lubricants and glidants
B   ❑ good flow behaviour of powders is critical to capsule filling
C   ❑ dissolution aids, wetting agents and super-disintegrants are sometimes added to capsules
D   ❑ there are no examples of enteric-coated capsules
E   ❑ capsules can be designed for non-oral routes of administration such as inhalation

Q96   Which of the following are types of softgel capsule used in the pharmaceutical industry?

1   ❑ meltable
2   ❑ chewable

3 ❏ suckable

    A ❏ 1

    B ❏ 2

    C ❏ 3

    D ❏ 1 and 3

    E ❏ 1, 2 and 3

Q97 Which of the following is not an advantage of softgel capsules?

    A ❏ improved drug absorption for poorly water-soluble drugs

    B ❏ better patient compliance when compared with other solid dosage forms

    C ❏ amenable to use with oil-soluble drugs

    D ❏ flexibility to deliver aqueous solutions of water-soluble drugs

    E ❏ protection against oxidative degradation by lipid vehicles

Q98 Which of the following is (are) in-process controls for rotary die encapsulation process for soft gelatin capsules?

    1 ❏ temperature

    2 ❏ timing

    3 ❏ pressure

    A ❏ 1

    B ❏ 2

    C ❏ 3

    D ❏ 1, 2 and 3

    E ❏ 1 and 3

Q99 Which of the following is false with reference to softgel formulations?

    A ❏ gelatin type B is the most commonly used material

    B ❏ glyercol is the most commonly used plasticiser

    C ❏ the water content in the final dry softgel is around 20%

    D ❏ titanium dioxide is normally used as an opacifier

    E ❏ synthetic or natural colours can be added to impart the desired colour to the shell

Q100 Which of the following best describes the types of softgel fill matrix?

    A ❏ lipophilic liquids

    B ❏ hydrophilic liquids

    C ❏ self-emulsifying oils

    D ❏ microemulsions

    E ❏ all of the above

Answers

A1   A

Magnesium chloride is soluble in 1 part water; the other three ingredients (B, C and D) are relatively insoluble, and methylcellulose forms a colloidal sol in hot water.

A2   D

Water has a high dielectric constant, has a negligible buffering capacity, is a non-selective solvent and is not suitable for extraction of organic matter.

A3   A

Miscibility refers to two liquids as opposed to solutes and a solvent. The solubility of a drug is the amount that moves into solution when equilibrium is reached.

A4   C

Co-solvents are vehicles used in combination to increase the solubility of a drug. They need to be miscible with the continuous phase. Weak electrolytes may require the addition of a co-solvent.

A5   E

Phytomenadione is vitamin K, an oil-soluble vitamin required for blood clotting and the formation of bones. It is also used as an antidote to bleeding, e.g. with warfarin doses that are too high. Being oil-soluble, it needs to be solubilised in an aqueous solution.

A6   D

Glycerol is viscous, has a sweet taste, is miscible with water and alcohol, and has preservative action (due to dehydration of microorganisms when used at a concentration of >20%). Cumulative amounts of ingested glycerol can lead to diarrhoea.

A7   D

Depending upon the solubility profile of the drug, it may get precipitated on dilution.

A8   A

As the length of a non-polar chain increases, the solubility of the compound in water decreases.

A9   E

A teaspoon is likely to lead to dose inaccuracy and therefore should be avoided, especially if a measuring device such as an oral syringe is available.

A10   B

Microbes thrive in an aqueous environment and the use of suitable preservatives is necessary.

A11   C

A reconstituted antibiotic forms a suspension in most cases; the other four are clear solutions.

A12   A

Ethanol is avoided in paediatric oral preparations because of its biological effects.

A13   D

Any improvement in solubility obtained by using hydrated form of a drug will not translate into improvement in dissolution rate.

A14   C

The drug concentration in the unstirred layer is usually higher than the diffusion layer. A change in environment around the diffusion layer, e.g. increased gastric contractions, will lead to an increase in drug solubilisation, with the possibility of better absorption.

A15   A

A solution contains dissolved solutes. Oil in a 60/40 ratio is not an aqueous solution.

A16   A

The HLB value is calculated for oil and water combinations such as emulsions and is not applicable to solutions.

A17   E

A solution is a single phase of two components and therefore it cannot contain a disperse phase. It consists of a solvent, which is usually but not necessarily the largest component.

A18   D

Pyrogen-free water is required for parenteral products; pyrogens, unlike most microorganisms, are not killed by heating with a bactericide.

A19   E

The sterility test is intended for detecting the presence of viable forms of microbes in pharmacopoeial preparations. Sterility testing of the products is carried out either by the membrane filtration method (Method A) or by the direct inoculation method (Method B). The technique of filtration is used whenever the nature of the product permits, i.e. for filterable aqueous preparations, for alcoholic or oily preparations, and for preparations miscible with, or soluble in, aqueous or oily solvents, provided these solvents do not have an antimicrobial effect in the conditions of the test. A membrane filter for sterility testing has a nominal pore size not greater than $0.45\,\mu\text{m}$.

A20   A

A rotary drum filter consists of a drum rotating in a tub of liquid to be filtered. The pressure difference between the outside of the drum compared to the inside pushes both liquid and solid onto its surface and the liquid is filtered through to the inside and pumped away. The solids adhering to the filter media of the drum are continuously removed by the knife to reveal a fresh media surface. Unlike other batch filtration processes where the process has to be stopped in order to remove filter cake, the rotary drum filter is a continuous process and filter cake is removed as the rotating drum advances the knife.

A21   B

Sintered glass filters are formed by fusion of glass granules. The porosity of filter depends upon the particle size of glass granules.

A22   D

In centrifuge sedimentors, the separation is due to the difference in the density of two or more phases. Centrifugal sedimentors are used for complete separation of solid–liquid mixtures and liquid–liquid mixtures. It also gives continuous discharge of two separated liquids and is widely used in emulsion separation.

A23   C

The objective of the filter aid is to prevent the filter medium from becoming blocked and to form an open, porous cake, hence reducing the resistance to flow of the filtrate. Talc is used as a filter aid in the process of filtration. It acts as an adsorbent and forms a fine surface deposit that screens out all solids, preventing them from contacting and plugging the supporting filter medium.

A24   E

The permeability coefficient or Darcy's constant, k, may be examined in terms of characteristics of the filter medium, such as porosity, specific surface area and compressibility. Thus, the permeability coefficient depends upon the nature of the filter medium.

A25   B

The rate of filtration is described by Darcy's law

$$\text{rate of filtration} = kA\Delta P/L\eta$$

where k = Darcy's constant, A = area of filter medium ΔP = pressure difference across the filter medium, L = thickness of filter medium and η = viscosity. According to the equation, the rate of filtration increases with an increase in pressure difference and area of filter, since the relation is directly proportional. However, the rate of filtration decreases with an increase in viscosity, since they are inversely related.

A26   D

One of the methods to validate the pore size of filters is the bubble point method. The test is performed by applying air pressure, or other gas pressure, to the filter in which the pores are filled with water. The pressure is gradually increased until bubbles pass through the filter and are detected. This bubble point pressure is inversely proportional to the diameter of the pores and thus is a measure of the largest pores.

A27   A

Removal or separation of a small amount of solid (< 1%) from the liquid is termed clarification. The term 'clarification' is applied when the solids do not exceed 1.0% and filtrate is the primary product.

A28   A

The flow of fluid through interstices of particles is expressed by Poiseuille's equation

$$dV/dT = AP/[\mu(\alpha W/A + R)]$$

where V = volume of filtrate, T = time, A = filter area, P = total pressure drop through cake and filter medium, μ = filtrate viscosity, α = average specific cake resistance, W = weight of dry cake solids and R = resistance of filter medium and filter.

A29   E

All the listed factors influence the bioavailability of drugs from aqueous suspension dosage forms. Despite the fact that the drug exists as insoluble suspended particles in the formulation, it still has to be dissolved in the gastrointestinal fluid before it is absorbed.

A30   D

Polysorbates (Tweens) and sorbitan esters (Spans) are surfactants used for oral administration. Sodium lauryl sulfate is used for external application.

A31   A

The choice of surfactant for parenteral administration is limited, the main ones used being polysorbates and some of the poloxamers and lecithins.

A32   D

If a suspension is deflocculated, the settling of the disperse particles will be slow. The slow rate of settling prevents the entrapment of liquid within the sediment, which will become very compacted, and lead to caking or claying. Aggregation of particles in a flocculated system will lead to a much faster rate of sedimentation. The sediment entraps a large amount of the liquid phase and can be redispersed by moderate agitation, but there is a danger of an inaccurate dose being administered. Therefore, it is more appropriate to have the suspension partially flocculated with viscosity controlled so that the sedimentation rate is at a minimum.

A33 C

Alginate mucilages must not be heated above 60 °C as polymerisation occurs, with a consequence of loss in viscosity. Alginates are most viscous immediately after preparation, after which there is a fall to a fairly constant value after about 24 h. They also have maximum viscosity over the pH range 5–9. At low pH, the acid will precipitate.

A34 C

Surfactants with HLB values in the range 7–9 are water-dispersible and would be suitable as wetting and spreading agents. Surfactants with HLB values below that range would be oil-soluble and suitable as water-in-oil emulsifying agents, whereas surfactants with HLB values above that range would be water-soluble and suitable as oil-in-water emulsifying agents.

A35 E

Methylcellulose is non-ionic. It is compatible with many ionic derivatives. It is stable over the pH range 3–11. On heating, the methylcellulose molecules will become progressively dehydrated and gel at 50 °C; on cooling, the original form is regained.

A36 B

Sodium carboxymethylcellulose is anionic and incompatible with polyvalent cations. It is stable over the pH range 5–10 and produces a clear solution in both hot and clear water.

A37 B

Large particles (diameter $> 5\,\mu m$) can give a gritty texture when applied externally.

A38 D

Both acacia and tragacanth are used internally, while bentonite is only used externally.

A39   D

Liquid paraffin and castor oil are formulated as emulsions for oral administration, while benzyl benzoate is formulated for external application.

A40   B

Cottonseed oil and safflower oil are used for their high caloric value in emulsion for intravenous administration, while cod liver oil, as a source of vitamins A and D, is used only orally.

A41   D

Non-ionic emulgents are usually less irritant and less toxic than the anionic and particularly the cationic emulgents. Ionic emulgents should not be used orally at the concentrations used for their emulsifying properties due to their irritancy on the gastrointestinal tract. Cationic emulgents are generally toxic, even at lower concentrations, and can only be used externally; some of them, e.g. cetrimide, are used as antiseptics.

A42   C

Theoretically, creaming could be totally prevented if the densities of the two phases were identical. In practice, however, it is almost impossible to have the two phases with identical densities. An increase in the density difference between the two phases would only increase the tendency of phase separation. The tendency of phase separation could be reduced by producing an emulsion with small droplet size and increasing the viscosity of the continuous phase. Emulsions containing less than 20% of the dispersed phase would be less stable. Therefore, it is desirable to have more than 20% of the dispersed phase proportion. However, phase inversion can occur when the dispersed phase exceeds 60%.

A43   B

Many of the oils and fats used in emulsions are susceptible to oxidation by atmospheric oxygen or by the action of microorganisms. Atmospheric oxidation can be minimised by adding antioxidants in the formulation, whereas oxidation of microbial origins can be controlled by the use of antimicrobial preservatives.

A44   E

Emulsions with a droplet size in the range of 1 nm to 1 μm could be transparent. Emulsions for intravenous administration must be of the oil-in-water type.

Water-in-oil emulsions are not dispersible in the aqueous environment and could cause embolism in the vein; therefore, they are not suitable for intravenous administration. Water-in-oil emulsions have an occlusive effect and will be effective for hydrating skin. Water-in-oil emulsions are more effective cleansers, especially for cleansing the skin of oil-soluble dirt, although their greasy texture may not be appealing to the patient.

A45   D

The HLB value of a combination of surfactants can be calculated as follows:

$$HLB = \frac{(\text{Quantity of surfactant 1} \times \text{HLB of surfactant 1}) + (\text{Quantity of surfactant 2} \times \text{HLB of surfactant 2})}{(\text{Quantity of surfactant 1} + \text{Quantity of surfactant 2})}$$

$$= \frac{(2 \times 4.3) + (3 \times 15)}{(2+3)}$$

$$= 10.72$$

A46   B

Let the amount of Span 80 in 5 g of the surfactant system be equal to y. Then:

$$\frac{(y \times 4.3) + (5-y) \times 15}{5} = 13$$

$$y = 0.93 \text{ g}$$

A47   E

Chloral hydrate is a hypnotic and sedative. The others are preservatives used in emulsions. The parahydroxybenzoic acid esters are probably the most commonly used group of preservatives. They are used at concentrations of 0.1–0.2% for both external and internal preparations.

A48   A

For the preparation of an acacia emulsion, the proportion of the amounts of oil, water and acacia in the primary emulsion should be 4:2:1. The oil and acacia are mixed thoroughly in a dry mortar. Then the designated amount of water is added to the primary emulsion all at once with rapid trituration, until a primary emulsion is formed. The primary emulsion is then diluted gradually with constant stirring to make up to the volume.

A49   E

The presence of water in creams means that preservatives are required to prevent microbial contamination.

A50   D

Water-miscible vehicles mix freely with water and therefore can be washed off the skin with ease.

A51   A

Wool fat and wool alcohol increase the water-holding capacity of greasy bases and form water-in-oil emulsions. Cetostearyl alcohol and emulsifying waxes both form oil-in-water emulsions.

A52   C

Ointments are topical preparations. These can be medicated or non-medicated and provide protective action.

A53   E

Cetomacrogol is non-ionic and produces oil-in-water creams.

A54   B

Cetrimide is cationic.

A55   A

Volatile propellants are used for aerosols, not for creams.

A56   B

Partitioning occurs and calculations need to be carried out for individual formulations.

A57   D

Pastes have up to 50% disperse phase and the increased consistency reduces spreadability.

A58   E

Fick's law is used to measure diffusion.

A59   B

Preservative should be soluble in the oily phase to ensure its availability throughout the formulation. A combination of preservatives is usually the best option to cover both phases.

A60   D

Water used in pharmaceutical products is expected to be free from microbial contamination. It is also a pharmacopoeial requirement.

A61   A

A pH electrode is used to measure and monitor pH changes over time; it does not affect stability.

A62   B

Because of lipid solubility and tendency to trap preservatives, higher concentrations are usually required in the case of ointments.

A63   C

The epidermis consists of a number of layers.

A64   D

Petroleum is also known as petrol, made from crude oil. Petrolatum is also known as petroleum jelly (Vaseline™).

A65   C

Compound hydroxybenzoate solution is a preservative solution.

A66   A

Topical application avoids the first-pass effect and can produce effects that are close to zero-order kinetics over prolonged time intervals. Drugs that are required for immediate onset are usually administered as parenteral products.

A67   D

Dilution of cream can reduce bioavailability; preservatives concentrations are usually halved on dilution.

A68   B

Most ointments are hydrophobic unless specially formulated with water-miscible bases.

A69   C

Most waxes melt around 60–70 °C. Heating to higher temperature is likely to cause oxidation of lipids and degradation of drugs. Therefore, ingredients should be heated to a temperature that is just sufficient to melt the ingredients and ensure emulsification in the liquid state.

A70   C

Transdermal preparations such as nicotine patches are designed for the systemic effects of the drug; these are not expected to provide any local action.

A71   B

The preparation of gels does not require heating. They are usually produced by cold mixing.

A72   D

Higher HLB values would mean higher solubility, leading to better water-removable characteristics.

A73   A

Hydrocarbon bases are the most occlusive and are therefore used in preparations such as lip-glosses. Absorption bases are second in protecting moisture loss.

A74   D

Tabletting technology is the most standardised, and the cost of manufacturing tablets is usually lower than that of other dosage forms.

A75   A

In all tabletting operations, the die is filled with the powder mix, compressed and the tablet ejected. The same principle is used in basic single-punch and high-speed rotary machines.

A76   E

Homogeneous mixture, good flow and compression, and addition of lubricants and glidants to achieve the desired adhesion properties are all important for tabletting.

A77   C

Fluidised-bed dryers are used for drying, not for size reduction.

A78   B

Because of the lower number of steps and excipients involved, direct compression tends to be the cheaper option.

A79   E

Polyvinyl pyrrolidone is used as a binder in tablets. If used at the high concentrations normally required for fillers, the tablets will be very hard and fail in disintegration tests.

A80   C

Magnesium stearate is universally accepted and used as a lubricant in the manufacturing of tablets and capsules.

A81   D

Polyvinyl pyrrolidone is useful as both a dry and a wet binder.

A82   E

Tablet disintegration can be achieved by any of the three mechanisms – swelling, water channelling due to surfactant, or effervescence by evolution of gas.

A83   D

Lubricants such as magnesium stearate tend to increase the disintegration time due to their hydrophobic effects.

A84 D

Buccal tablets can be designed to be bioadhesive so they stick to the buccal mucosa and release drug over a period of time. Up to 4 h of adhesion and release can be achieved easily by buccal tablets. All other listed tablets are for immediate release.

A85 C

Lozenges do not contain disintegrants.

A86 B

Vaginal tablets are the best option for rapid release and stability in tropical conditions. Suppositories melt at temperatures higher than body temperature, and other dosage forms tend to be costlier.

A87 E

Tablets can be designed to be non-irritant and rapidly disintegrating, leading to a bioadhesive solution that is retained for a prolonged period.

A88 A

Polymethacrylates (Eudragits) are soluble in water and non-aqueous solvents. All other listed compounds are only water-soluble.

A89 D

Prolonged-release tablets can be designed based on any of the three mechanisms.

A90 C

Release profiles are determined by dissolution, not by disintegration testing.

A91 D

Dissolution studies help in matching release profiles required during the development of generic formulations. They are also a requirement for quality control.

A92   A

To prepare hard gelatin shells, the pins are dipped in a concentrated solution of gelatin, plasticiser, colour and other additives. These are spun, dried, stripped and cut to the required size. Two pieces are then joined.

A93   A

The size of capsules and fill volumes are inversely linked: the smaller the size, the larger the fill volume.

A94   D

The capsule shells are made of gelatin, plasticiser and water. Because of their chemical nature, aqueous solutions cannot be filled in the capsules.

A95   D

Capsules can be coated to provide enteric protection. Hydroxypropyl-methylcellulose (HPMC) capsules are generally used for such purposes.

A96   E

Softgel capsules used in pharmaceuticals can be meltable, chewable and suckable.

A97   D

Similar to Q94, water-soluble drugs can be loaded in capsules as aqueous solutions.

A98   D

The preparation of softgels is based on heating, and therefore temperature, timing and pressure are critical.

A99   C

The final water content in capsules is 5–7%.

A100   E

The softgel matrix can be of any kind, as long as it has no or only very small proportions of water. Suspensions, lipids, nanoemulsions and microemulsions can easily be encapsulated.

Test 5

Advanced drug-delivery systems

Beverley D Glass, Alison Haywood,

Roop K Khar and Sanjay Garg

Introduction

Over the past decade, despite the importance of the biotechnology industry and increasing expenditure in research and development, there have been only a limited number of drugs originating from large pharmaceutical companies. The pharmaceutical industry has turned its attention to advanced drug-delivery systems such as pulmonary, transdermal, rectal and vaginal, proteins and gene, nasal, buccal and ophthalmic, parenteral, total parenteral nutrition (TPN) and aerosol delivery, and these are the focus of this test. The philosophy behind these delivery systems is to increase therapeutic effects while decreasing toxicity by increasing the drug in the vicinity of the target cells and reducing drug exposure to non-target cells. The benefits of this approach include improved patient adherence and acceptance of their medications, improved outcomes and a reduction in adverse effects. The medications are such that they may be given to outpatients, resulting in a reduction in the overall use of medical resources.

Also included in this test is the concept of intellectual property (IP). There have been claims, especially related to drugs that treat acquired immunodeficiency syndrome (AIDS), that patents in the pharmaceutical industry have deprived people of access to life-saving drugs because of high prices. This has been refuted by the fact that many drug companies assist developing countries through a variety of innovative public–private partnerships to ensure that drugs reach those people in need.

Pulmonary delivery administers drugs directly to the respiratory tract, resulting in a rapid onset of action and the potential for the delivery of smaller doses locally compared with oral and parenteral delivery. This reduces adverse systemic effects and drug costs. Transdermal delivery involves manipulating the physicochemical properties of drugs to penetrate the skin, resulting in systemic absorption; the ultimate objective is to incorporate suitable drugs into either vehicles or devices such as patches for delivery to an active site in a controlled manner.

When the oral route is not suitable due to problems with the gastrointestinal tract, or when local effects are desired, drugs may be administered rectally or vaginally.

Although biotechnology products account for most of the proteins used in therapy or under development, e.g. human insulin, erythropoietin and monoclonal antibodies, there are still important proteins that are derived from the blood of humans or animals, such as albumin. Although the nasal route has been used for topical administration for years, it has been investigated as a route for systemic delivery, especially for proteins and peptides. The buccal route has also been investigated as an alternative route of administration for biopharmaceuticals. Advantages include accessibility and avoidance of hepatic first-pass effects; however, there is still low bioavailability of proteins.

Although ophthalmic preparations are presented as a diversity of formulations, including drops, lotions, ointments and gels, they are all required to be sterile for administration to the eye. This special requirement for sterility is continued to parenterals, which, when formulated as solutions, are also required to be particle-free for injection or infusion into the body. Aerosols, colloidal dispersions of liquids or solids in gases, although representing a dosage form particularly suited to delivery of drugs to the lungs, are also well suited for delivering topical preparations.

The final section of this test is dedicated to IP and provides important information for the pharmacist relating both to new chemical entities on the market and to generic drugs.

Questions

Q1 The most important mechanism for particle deposition in the small airways and alveoli for particles in the size range 0.5–3 µm is:

A ❏ interception
B ❏ gravitation sedimentation
C ❏ electrostatic interactions
D ❏ inertial impaction
E ❏ Brownian diffusion

Q2 Particles with a diameter of approximately 0.5 µm are:

A ❏ too large to be deposited via gravitational sedimentation
B ❏ too small to be deposited via Brownian diffusion
C ❏ too large to be deposited via inertial impaction
D ❏ immediately exhaled since they are not deposited by any mechanism
E ❏ unaffected by hygroscopic growth

Questions 3 and 4 involve the following case:

A pharmaceutical manufacturer that is currently developing a new steroid aerosol has found that during early clinical trials most patients were experiencing oral candidiasis as a side-effect. The steroid is lipophilic and has an approximate particle size diameter of 8 µm. You have been asked to provide potential solutions to this problem.

Q3 Which of the following recommendations is most likely to reduce the side-effect of oral candidiasis, while maintaining the steroid's efficacy?

A ❏ decrease the dose of the active ingredient
B ❏ increase the hydrophilicity of the active ingredient
C ❏ decrease the particle size of the active ingredient
D ❏ redesign the aerosol device
E ❏ increase the particle size of the excipients

Q4 Which of the following statements is true when considering the effects of hygroscopic growth on the above active ingredient (steroid)?

A ❏ it is due to condensation of water on the particle surface due to a decrease in relative humidity
B ❏ it is due to the increased vapour pressure of the solution on the particle surface

C ❏ it will result in a decrease in particle size
D ❏ it will result in deposition lower in the respiratory tract than originally predicted
E ❏ it is negligible since the active ingredient is lipophilic

Q5 Sodium cromoglycate can be effectively delivered via the pulmonary route because:

1 ❏ it is poorly absorbed orally
2 ❏ the abundance of capillaries and large surface area of above 60 m^2 allows for good systemic activity
3 ❏ the lung has no metabolic activity and first-pass metabolism in the liver is avoided
  A ❏ 1, 2 and 3
  B ❏ 1 and 2
  C ❏ 2 and 3
  D ❏ 1
  E ❏ 3

Q6 When considering the propellants used in metered dose inhalers (MDIs), hydrofluoroalkanes (HFAs):

A ❏ contain excipients that may increase the droplet size of the emitted aerosols
B ❏ have higher boiling points than chlorofluorocarbons (CFCs) such as CFC-11 and CFC-114
C ❏ are good solvents for the surfactants commonly used in MDIs
D ❏ are more toxic than hydrocarbon propellants
E ❏ do not contribute to global warming

Q7 When comparing metered dose inhalers (MDIs) with dry powder inhalers (DPIs):

A ❏ much of the active ingredient is lost through impaction in MDIs since particles exit at a velocity greater than 30 m s^{-1}
B ❏ MDIs are not suitable for excipients that are sensitive to oxidation
C ❏ MDIs have an increased risk of microbial contamination during cold filling
D ❏ the efficiency of MDIs is affected by relative humidity
E ❏ propellants used in MDIs are good solvents for most active ingredients

Q8   A patient has been experiencing bronchoconstriction when using their jet nebuliser. Which of the following recommendation(s) would you make?

1   ❏ consider using an ultrasonic nebuliser as opposed to a jet nebuliser
2   ❏ consider using a unit dose nebuliser formulation as opposed to a multidose preparation
3   ❏ consider diluting the nebuliser formulation with sterile saline solution
A ❏ 1, 2 and 3
B ❏ 1 and 2
C ❏ 2 and 3
D ❏ 1
E ❏ 3

Q9   With reference to jet nebulisers, which of the following statements is true?

A   ❏ the mass median aerodynamic diameter (MMAD) increases as the flow rate of gas is increased
B   ❏ the rate of drug delivery is inversely proportional to the rate of gas flow
C   ❏ the inspiratory phase of breathing constitutes one-third of the breathing cycle
D   ❏ aerosol production is always independent of the patient's tidal volume
E   ❏ a larger therapeutic dose can be delivered via a dry powder inhaler (DPI) compared with a jet nebuliser

Q10   When comparing ultrasonic and vibrating-mesh nebulisers, which of the following statements is true?

A   ❏ piezoelectric crystals are utilised only in ultrasonic nebulisers
B   ❏ vibrating-mesh nebulisers require more patient inhalation coordination than ultrasonic nebulisers
C   ❏ vibrating-mesh nebulisers result in a greater increase in solution temperature compared with ultrasonic nebulisers
D   ❏ drug solubility is likely to decrease when using ultrasonic nebulisers
E   ❏ vibrating-mesh nebulisers have a smaller residual volume than ultrasonic nebulisers

Q11   Which of the following is true for nebuliser fluid formulations?

A   ❑ ultrasonic nebulisers show improved delivery of suspensions compared with jet nebulisers

B   ❑ sulfites are the preferred antioxidants for use in nebuliser fluids

C   ❑ ultrasonic nebulisers are useful in delivering recombinant human deoxyribonuclease for the treatment of cystic fibrosis

D   ❑ jet nebulisers may increase the surface tension and viscosity of nebuliser fluids

E   ❑ nebuliser solutions should have a pH lower than 5

Questions 12–15 involve the following case:

Fentanyl (molecular weight 336; solubility in water 1 in 10 000) is an opioid analgesic commonly used in the treatment of acute pain, for chronic pain for patients intolerant of morphine, and for breakthrough pain for patients with cancer stabilised on an opioid analgesic.

Q12   Which of the following is not a reason to justify a transdermal delivery system for fentanyl?

A   ❑ its solubility in water

B   ❑ the molecule is un-ionised

C   ❑ its molecular weight

D   ❑ its potency

E   ❑ its therapeutic index

Q13   Penetration of the stratum corneum by fentanyl:

A   ❑ need not occur for the drug to act systemically

B   ❑ proceeds by an active diffusion process

C   ❑ is lowered when the skin is damaged or abraded

D   ❑ is not affected by the vehicle in which it is applied

E   ❑ is proportional to the drug concentration and the partition coefficient of the drug between the tissue and the vehicle

Q14   A form of Fick's law describes steady-state transport through the skin:

$$J = (DP/\delta)\Delta Cv$$

Which of the following statements is incorrect?

A   ❑ drug flux involving bidirectional movement of the drug at the membrane or surface is possible

B   ❑ the larger the drug diffusion coefficient, the greater the drug flux into the skin

C ❏ the smaller the drug partition coefficient between the vehicle and the skin, the greater the drug flux into the skin

D ❏ the greater the thickness of the stratum corneum, the lower the drug flux

E ❏ the greater the difference in drug concentration between the vehicle and the tissue, the greater the drug flux

Q15 Which of the following physicochemical factors would result in improved penetration of the stratum corneum during transdermal delivery of fentanyl?

A ❏ increased hydration
B ❏ decreased temperature
C ❏ decreased pH
D ❏ decreased drug concentration
E ❏ increased molecular size of the drug

Q16 Which of the following statements about transdermal drug delivery is (are) false?

1 ❏ transport of water-soluble drugs is through endogenous lipids within the stratum corneum, the bulk of this being intercellular

2 ❏ if the drug is more soluble in the stratum corneum than the vehicle in which it is presented, the concentration in the first layer of the skin may be lower than that in the vehicle

3 ❏ if depletion of the contact layer of the vehicle occurs, then the nature of the formulation will dictate how readily these are replenished by diffusion and therefore will dictate the rate of absorption

A ❏ 1, 2 and 3
B ❏ 1 and 2
C ❏ 2 and 3
D ❏ 1
E ❏ 2

Q17 Although manufacturers design transdermal patches in a variety of ways, they may be categorised as either monolith (matrix) or rate-limiting membrane systems. Which of the following statements about the membrane-controlled system is true?

A ❏ the drug matrix (polymeric material) is between the backing and frontal layers

B ❏ the polymeric material controls the rate of release of the drug

C   ❏ a matrix may be of two types – with or without excess in respect of the equilibrium solubility and steady-state concentration gradient at the stratum corneum

D   ❏ a drug reservoir or pouch in a liquid or gel form is contained within a rate-controlling membrane and backing, adhesive and protecting layers

E   ❏ in these patches, the Higuchi square root of time law is usually obeyed

Q18   Which of the following is not an advantage of transdermal delivery?

A   ❏ it avoids gastrointestinal drug absorption difficulties
B   ❏ it avoids the first-pass effect
C   ❏ it allows extended therapy, thus improving compliance
D   ❏ therapy can be terminated rapidly
E   ❏ it has a rapid onset of action

Q19   Transdermal drug delivery may be enhanced by iontophoresis. Which of the following statements about iontophoresis is incorrect?

A   ❏ iontophoresis is an electrochemical method using an applied electrical current
B   ❏ iontophoresis enhances the transport of charged drug molecules across the skin
C   ❏ the process involves electrostatic repulsion at the electrode
D   ❏ negative ions are delivered by the cathode, while positive ions are delivered by the anode
E   ❏ combinations of iontophoresis and other penetration enhancers do not result in synergy and fewer side-effects

Q20   Which of the following drugs would be suitable for iontophoretic transdermal drug delivery?

A   ❏ fentanyl citrate
B   ❏ morphine
C   ❏ indometacin
D   ❏ diclofenac
E   ❏ ketoprofen

Q21 In the following transdermal gel matrix formulation, which of the formulation components controls the release of the drug across the skin?

A ❑ estradiol
B ❑ carbomer 934P
C ❑ triethanolamine
D ❑ propylene glycol
E ❑ ethanol

Q22 In the case of suppositories, which of the following statements is true?

A ❑ suppositories are semisolid dosage forms, intended for insertion into body orifices, where they melt, soften or dissolve
B ❑ suppositories may exert local action to relieve constipation or the pain, irritation and inflammation associated with haemorrhoids and anorectal conditions
C ❑ suppository bases such as theobroma oil and macrogols do not influence the release of the drug incorporated into the base
D ❑ glycerin suppositories do not promote laxation, as they do not have a dehydrating effect on the mucous membranes
E ❑ suppositories may not exert systemic action, since the mucous membranes of the rectum do not permit the absorption of drugs

Q23 A limitation in using suppositories, compared with orally administered drugs, is that drugs:

A ❑ that are destroyed by pH or the enzymatic activity of the stomach may be administered in this way
B ❑ bypass the liver after rectal absorption
C ❑ administered rectally are used in the treatment of patients with vomiting episodes
D ❑ may be administered to adult or paediatric patients who are unable to swallow
E ❑ may be slowly or incompletely absorbed, with inter- and intra-subject variations reported

Questions 24 and 25 involve the following case:

There are a number of physiological factors affecting rectal absorption, including colonic content, the circulation route, and the fact that rectal fluids are essentially neutral, resulting in the form in which the drug is administered not being chemically changed by the rectal environment.

Q24 Which of the following drugs should not be incorporated in a fatty suppository base such as cocoa butter?

A   ❑ prochlorperazine
B   ❑ indometacin sodium
C   ❑ oxymorphone hydrochloride
D   ❑ ergotamine maleate
E   ❑ chlorpromazine hydrochloride

Q25 Which of the following are physicochemical factors of drugs and suppository bases that are important in the formulation of suppositories?

1   ❑ a lipid/water partition coefficient of the drug is an important consideration in the selection of a suppository base
2   ❑ water-soluble bases release for absorption both water-soluble and oil-soluble drugs
3   ❑ for drugs in the suppository in the undissolved state, the size of the drug particles will not influence the rate of dissolution

A ❑ 1, 2 and 3
B ❑ 2 and 3
C ❑ 1 and 2
D ❑ 3
E ❑ 2

Q26 Which of the following is a disadvantage of using synthetic fats over natural fatty bases, such as theobroma, as suppository bases?

A   ❑ solidifying points are unaffected by overheating
B   ❑ these bases offer resistance to oxidation
C   ❑ different melting point ranges allow for varying degrees of hardness
D   ❑ fluidity is greater than for theobroma, which may necessitate the use of thickeners to reduce the rate of sedimentation
E   ❑ no lubricant is necessary, as they contact on cooling

Q27 An ideal suppository base:

1   ❑ melts at room temperature and dissolves or disperses in body fluids
2   ❑ retains its shape on handling and is stable at body temperature

3 ❏ is non-toxic and non-irritant to the mucous membranes of the rectum

    A ❏ 1, 2 and 3

    B ❏ 1 and 2

    C ❏ 2 and 3

    D ❏ 3

    E ❏ 1 and 3

Q28 Macrogols (polyethylene glycols) are commonly used as suppository bases because:

A ❏ they can be used as mixtures of two or more grades to achieve a melting point above 42 °C

B ❏ they absorb water well and are hygroscopic

C ❏ they have good solvent properties, which results in retention of the drug in the base in the body

D ❏ they sometimes fracture on storage

E ❏ they may promote crystal growth of the drug within the suppository

Q29 Although the vaginal route is commonly used for the local effects of drugs, it is also a useful route for the systemic administration of drugs. Which of the following statements is correct?

A ❏ bioavailability is improved due to the drug not bypassing the first-pass effect

B ❏ the vaginal wall, although suited to absorption, does not have a vast network of blood vessels

C ❏ many drugs may be administered by this route

D ❏ preparations come into contact with tissues in the vagina that are prone to infection and therefore need to be free of offending organisms

E ❏ vaginal gels are not preserved with antimicrobial agents

Q30 Some vaginal inserts are capsules of gelatin containing a medicament to be released intravaginally. Which of the following statements is incorrect?

A ❏ these inserts are made of gelatin to which glycerin or a polyhydric alcohol such as sorbitol has been added in order to render the gelatin elastic- or plastic-like

B ❏ these inserts contain less moisture than hard capsules

C ❏ these inserts are preserved by methyl or propyl paraben

D   ❑ these inserts may be used to encapsulate liquids, suspensions, pasty materials and dry powders

E   ❑ these inserts are elegant and easily applied by the patient

Q31 Vaginal inserts/tablets may be used more widely than pessaries. Which of the following statements is incorrect?

A   ❑ these tablets are easier to manufacture

B   ❑ these tablets are more stable and less messy

C   ❑ these tablets may be prepared by standard tablet compression

D   ❑ these tablets include standard excipients such as fillers, disintegrating agents, dispersing agents and lubricants

E   ❑ excipients may need to be added in order to improve the disintegration of these tablets due to the high moisture content of the vagina

Questions 32 and 33 involve the following case:

Proteins are built up of amino acids. Their diffusional transport through epithelial barriers in the gastrointestinal tract is slow. Because conditions in the lumen of the gastrointestinal tract are extremely hostile to proteins, enzymatic degradation is usually fast. This therefore requires the majority of pharmaceutical proteins to be delivered via the parenteral route. You have been asked to provide potential solutions to this problem.

Q32 Pharmaceutical proteins clearly offer a special challenge to the formulation scientist. Which of the following statements is incorrect?

A   ❑ proteins are delicate, small molecules with many functional groups

B   ❑ because the three-dimensional structure is stabilised by weak physical bonds, it can be changed readily and irreversibly

C   ❑ alteration of this three-dimensional structure can affect both the interaction at the receptor and the pharmacokinetic parameters

D   ❑ changes in this three-dimensional structure can make the protein immunogenic

E   ❑ transporter molecules are required to facilitate epithelial penetration of proteins

Q33 Degradation rates of proteins depend on environmental conditions, and the formulation pharmacist should be careful in selecting conditions for optimal stability. Physical instability can be caused by:

1   ❑ elevated temperatures
2   ❑ low temperatures
3   ❑ shaking/exposure to shear forces
    A   ❑ 1 and 3
    B   ❑ 2 and 3
    C   ❑ 1
    D   ❑ 3
    E   ❑ 1, 2 and 3

Q34 Which of the following excipients should not be used in the formulation of pharmaceutical proteins?

A   ❑ amino acids
B   ❑ detergents
C   ❑ phosphate
D   ❑ reducing sugars
E   ❑ ascorbic acid

Q35 Conserving the integrity of pharmaceutical proteins is essential to ensure the optimal therapeutic effect and to minimise, for example, the induction of an immune response. Which of the following statements about the shelf life and storage of pharmaceutical proteins is incorrect?

A   ❑ most proteins can be stabilised by refrigeration
B   ❑ it is preferred that pharmaceutical proteins are stored in the dry form and reconstituted
C   ❑ the three-dimensional delicate structures of pharmaceutical proteins are preferentially dried by freeze drying
D   ❑ the choice of proper excipients such as lyoprotectants for this freeze drying stage is extremely important
E   ❑ the preferred shelf life for pharmaceutical products is a minimum of 2 years, which is difficult to achieve for aqueous solutions of pharmaceutical proteins

Q36 Which of the following statements about the chemical degradation of proteins is incorrect?

A   ❑ deamidation, a common degradation reaction in water, is dependent on pH and the neighbouring amino acids
B   ❑ oxidation is catalysed by traces of transition metals

C ❑ under neutral conditions, most of the peptide bonds between amino acids are stable

D ❑ an oxidative environment may cause free cysteine units to form disulfide bridges

E ❑ isomerisation of the naturally occurring D-form to the L-form will change the structure of the protein

Q37 During freeze drying of proteins, water is removed by sublimation. A lyoprotectant is necessary to stabilise the product as the removal of water may irreversibly affect the protein structure. Which of the following statements about lyoprotectants is incorrect?

A ❑ lyoprotectants are reducing sugars and readily form reconstitutable porous cakes

B ❑ lyoprotectants replace water as a stabilising agent of the protein

C ❑ lyoprotectants increase the glass transition temperature, thus avoiding collapse of the porous cake

D ❑ lyoprotectants slow down the removal of water from the frozen cake, which would interfere with the rapid constitution of the freeze-dried cake

E ❑ lyoprotectants slow down secondary drying processes and minimise the chances of overdrying

Q38 Finding alternatives to the parenteral route for the delivery of proteins has been a challenge for the pharmaceutical scientist. Which of the following administration routes does not result in low bioavailability of proteins?

A ❑ nasal

B ❑ pulmonary

C ❑ rectal

D ❑ buccal

E ❑ transdermal

Q39 Which of the following techniques represents an immunological approach to determining protein structure?

A ❑ mass spectrometry

B ❑ fluorimetry

C ❑ enzyme-linked immunosorbent assay (ELISA)

D ❑ affinity chromatography

E ❑ ion-exchange chromatography

Questions 40 and 41 involve the following case:

Pharmaceutical preparations are applied topically to the eye to treat surface or intraocular conditions such as bacterial infections, allergic conjunctivitis, dry eye and glaucoma. In conditions such as glaucoma, both systemic drug use and topical treatments may be employed.

Q40 Ocular bioavailability is an important factor in the effectiveness of an applied medication. Physiological factors that affect ocular bioavailability are protein binding, drug metabolism and lacrimal drainage. Which of the following statements about ocular bioavailability is (are) correct?

1 ❑ protein-bound drugs are incapable of penetrating the corneal epithelium because of the size of the protein–drug complex

2 ❑ although tears may contain enzymes such as lysozyme, these are incapable of metabolic degradation of drug substances

3 ❑ ophthalmic suspensions, gels and ointments mix readily with the lacrimal fluids, causing them to remain longer in the cul-de-sac and enhancing drug activity

A ❑ 1, 2 and 3
B ❑ 1 and 2
C ❑ 1 and 3
D ❑ 1
E ❑ 2

Q41 Which of the following statements about ophthalmic drug delivery is incorrect?

A ❑ the normal volume of tear fluid in the cul-de-sac of the human eye is about 7–8 μL

B ❑ a single drop of an ophthalmic solution is about 50 μL, so much of this administered drop is lost

C ❑ since retention of an ophthalmic solution on the eye surface is significant, a large fraction of the drug administered is absorbed

D ❑ patients often have to repeatedly administer ophthalmic solutions to the eye, which may result in decreased patient adherence

E ❑ decreased frequency of dosing may be achieved by formulations such as gels, ophthalmic suspensions and ointments

Q42 Ophthalmic solutions need to be sterilised for safe use. Although it is preferable to sterilise in the final container by autoclaving, thermal instability of the drug might preclude this. What alternative method of sterilisation is recommended, especially for use in a practice setting?

A ❏ dry heat
B ❏ gas
C ❏ filtration
D ❏ ionising radiation
E ❏ ultraviolet light

Q43 The pH of an ophthalmic preparation may be adjusted and buffered. Which of the following reasons for buffering is incorrect?

A ❏ buffering provides formulation stability
B ❏ buffering enhances lipid solubility
C ❏ buffering proves comfortable to the eye
D ❏ buffering enhances bioavailability
E ❏ buffering maximises preservative efficacy

Q44 Which of the following formulation considerations need to be taken into account when preparing a product for ophthalmic delivery?

A ❏ solutions need not be clear and free of all particulate matter in order that the product is aesthetically pleasing
B ❏ an ophthalmic suspension may be formulated when extended corneal time is not required or when the drug is soluble in an aqueous vehicle
C ❏ drug particles in an ophthalmic suspension need not be finely subdivided in order to avoid irritation
D ❏ the suspended particles must not aggregate and must redisperse easily on shaking
E ❏ ointment bases are required to be non-irritant and soften at a point close to room temperature

Q45 Although the nasal cavity has typically been the route of administration for topical delivery of decongestants, the limitation of nasal administration of drugs for systemic absorption is:

A ❏ convenience
B ❏ a good area for absorbing drugs
C ❏ a sufficient blood supply

D ❑ the presence of cilia, which facilitate the movement of mucus, resulting in mucociliary clearance, which is a non-specific defence action

E ❑ the nasal cavity is lined with mucous membrane

Q46 Which of the following statements about nasal drug delivery is incorrect?

A ❑ mucus in the nasal cavity presents a diffusional barrier to drug absorption

B ❑ when formulating drugs for nasal delivery, it is important that the interruption of the mucociliary clearance from the drug or excipients is minimal or temporary

C ❑ formulation of drugs to overcome the diffusional barrier needs to be such that the drug remains in contact long enough to be released and absorbed

D ❑ the nasal mucosa possess enzymatic activity offering protection against exogenous chemicals

E ❑ nasal first-pass metabolism is an insignificant factor on the absorption of all drugs

Q47 Which of the following components is the preservative in a nasal spray?

A ❑ chlorobutanol

B ❑ phenylephrine hydrochloride

C ❑ sodium metabisulfite

D ❑ sodium chloride

E ❑ propylene glycol

Q48 A physiochemical factor of the drug that does not affect nasal absorption is:

A ❑ the molecular weight and size of the drug

B ❑ the pH and lipid partition coefficient

C ❑ the drug class

D ❑ the solubility of the drug

E ❑ the powder morphology and particle size

Q49 Absorption enhancers increase the rate at which drugs pass through the nasal mucosa. Which of the following enhancers may not be suitable because it (they) cause(s) mucosal damage?

1 ❑ surfactants

2 ❑ bile salts

3   ❑ cyclodextrins

A   ❑ 1, 2 and 3
B   ❑ 1 and 2
C   ❑ 2 and 3
D   ❑ 3
E   ❑ 1

Q50   Metered-dose pump systems are an important nasal dosage form. Which of the following statements about these dosage forms is incorrect?

A   ❑ these dosage forms may be formulated as solutions, suspensions or emulsions
B   ❑ they deliver a predetermined volume of 25–200 µL
C   ❑ deposition occurs over a small surface area
D   ❑ particle size is important
E   ❑ the volume delivered is affected by the size of the nasal cavity

Q51   Buccal tablets are flat, oval tablets intended to be dissolved in the buccal cavity. Which of the following statements about buccal drug delivery is incorrect?

A   ❑ buccal tablets are designed for absorption through the oral mucosa
B   ❑ buccal tablets are designed to erode rapidly
C   ❑ buccal tablets enable the oral absorption of drugs destroyed in the gastrointestinal tract
D   ❑ buccal tablets enable the oral absorption of drugs poorly absorbed from the gastrointestinal tract
E   ❑ lozenges and troches, which are often prepared extemporaneously, are examples of solid dosage forms other than tablets suitable for buccal delivery

Q52   Which of the following advantages are offered by parenterally administered drugs, when compared with other administration routes?

A   ❑ rapid onset of action
B   ❑ predictable effect
C   ❑ avoid the gastrointestinal tract
D   ❑ reliable in very ill patients
E   ❑ all of the above

Q53 Which of the following is a limitation of parenteral preparations?

A ❏ pain at the site of injection
B ❏ higher cost compared with other routes of administration
C ❏ needle fear
D ❏ difficulty of fixing an incorrect dose
E ❏ all of the above

Q54 Identify the incorrect statement from the following about intramuscular injections:

A ❏ compared with subcutaneous injections, there is a better vascular supply and therefore more rapid absorption
B ❏ the maximum volume for intramuscular injection is 4 mL
C ❏ dissolution and absorption can be controlled to produce long-acting products
D ❏ bioavailability is 100%
E ❏ molecular weight of the drug affects the fraction cleared

Q55 Parenteral products can be formulated as:

A ❏ suspensions
B ❏ solutions
C ❏ emulsions
D ❏ lotions
E ❏ A, B and C

Q56 Which of the following is correct with reference to the effect of calcium on the stability of intravenous lipid emulsions?

A ❏ improves stability by neutralising the negative surface of lipid droplets
B ❏ limits stability by neutralising the negative surface of lipid droplets
C ❏ improves stability through pH-buffering action
D ❏ limits stability through pH-buffering action
E ❏ calcium has no effect on intravenous lipid emulsion stability

Q57 After intravenous administration, the maximum concentration for most drugs is achieved in:

A ❏ 5 s
B ❏ 45 s
C ❏ 4 min
D ❏ 10 min
E ❏ 1 h

Q58 Select the correct statement with reference to large-volume parenterals (LVPs):

A ❑ LVPs are not buffered
B ❑ phosphoric acid buffer can be used to buffer LVPs
C ❑ buffers can be used in LVPs, provided the mixture is isotonic
D ❑ LVPs can have up to 5% suspended particles
E ❑ LVPs can be hypertonic because they are injected over a long period of time

Q59 Which of the following is not related to parenteral solutions?

A ❑ clear
B ❑ sterile
C ❑ pyrogen-free
D ❑ no dissolved O_2
E ❑ packed to ensure sterility

Q60 Electrolytes are added in parenteral preparations in order to:

A ❑ prevent water loss
B ❑ adjust pH and tonicity
C ❑ prevent microbial growth
D ❑ improve appearance
E ❑ prevent precipitation

Q61 The maximum volume (in mL) that can be injected by the subcutaneous route is:

A ❑ 0.2
B ❑ 1.0
C ❑ 10.0
D ❑ 100.0
E ❑ 200.0

Q62 The duration of action from an intravenous preparation depends upon:

1 ❑ dose
2 ❑ timescale of administration (bolus or infusion)
3 ❑ distribution, metabolism and excretion
  A ❑ 1
  B ❑ 2
  C ❑ 3
  D ❑ 1 and 3
  E ❑ 1, 2 and 3

Q63 Which one of the following is not true for intramuscular injections?

A ❑ the pH of muscle tissue is more acidic than the pH of most physiological fluids
B ❑ high pH may cause precipitation of some drugs
C ❑ hydrophobic drugs may bind to muscle proteins
D ❑ the intramuscular route can be used for prolonged action
E ❑ molecules diffuse through muscle fibres via the pores of capillary walls to the blood

Q64 Which one of the following cannot be used as non-aqueous or mixed solvent in parenteral preparations?

A ❑ glycerol
B ❑ ethanol
C ❑ propylene glycol
D ❑ PEG 300
E ❑ methylene chloride

Q65 Sterile filtration of parenteral preparations is carried out by using filters of the following pore size (in µm):

A ❑ 0.45
B ❑ 0.22
C ❑ 0.66
D ❑ 1.0
E ❑ 1.22

Q66 The size range (in µm) for lipid globules in oil-in-water emulsions used for parenteral nutrition is:

A ❑ 0.01–0.1
B ❑ 0.2–0.8
C ❑ 0.9–2.0
D ❑ 2.1–5.0
E ❑ 10.0–50.0

Q67 Which of the following factors can adversely affect stability of an intravenous lipid system?

A ❑ emulsifier
B ❑ surfactants
C ❑ antioxidants
D ❑ light protection
E ❑ added trace elements

Q68 Oxygen in total parenteral nutrition products can influence stability and shelf life of oxidisable drugs such as vitamins, lipids and amino acids. Which of the following is (are) correct option(s) for the source of oxygen in these products?

1 ❑ infusions and additives
2 ❑ aeration during fluid transfers
3 ❑ residual headspace and permeation in the bag
A ❑ 1
B ❑ 2
C ❑ 3
D ❑ 1 and 2
E ❑ 1, 2 and 3

Q69 A major issue with parenteral therapy is thrombophlebitis. Which of the following does not help in reducing thrombophlebitis?

A ❑ minimum possible particulates
B ❑ smallest possible duration of injection
C ❑ large cannula size
D ❑ correct pH and osmolarity
E ❑ use of polyurethane tubing

Q70 Which of the following is correct as a possible source of particulate contamination in parenteral preparations?

A ❑ active and excipients
B ❑ packaging
C ❑ people
D ❑ environment
E ❑ all of the above

Q71 Which of the following is (are) correct with reference to the potential hazards of parenteral formulations?

1 ❑ overloading circulatory system
2 ❑ allergic reaction
3 ❑ venous thrombosis/thrombophlebitis
A ❑ 1
B ❑ 2
C ❑ 3
D ❑ 1 and 2
E ❑ 1, 2 and 3

Q72 Which of the following is not true for intrathecal administration?

A ❏ administration of drugs in solution is done through an intrathecal catheter
B ❏ it offers an opportunity to deliver drugs to the brain and spinal cord
C ❏ it is frequently used
D ❏ it is more invasive than intravenous, intramuscular and subcutaneous administration
E ❏ implanted catheters and pumps have been used to reduce the risk of infection on repeated puncture

Q73 Which of the following is (are) true with reference to drugs and dosage-form-related factors influencing parenteral drug absorption?

1 ❏ pK_a, solubility, dissolution rate, diffusivity, partition coefficient and drug concentration
2 ❏ types of dosage form and injection volume
3 ❏ physicochemical properties of the dosage form
A ❏ 1
B ❏ 2
C ❏ 3
D ❏ 1 and 2
E ❏ 1, 2 and 3

Q74 Which of the following is not a type of parenteral controlled release system?

A ❏ dissolution-controlled depot formulations
B ❏ adsorption-type depot formulations
C ❏ concentrated suspension with drug depot for intravenous administration
D ❏ esterification-type depot formulations
E ❏ encapsulation-type depot formulations

Q75 Which of the following is (are) an example(s) of chemical modifications to achieve controlled release?

1 ❏ penicillin G potassium
2 ❏ procaine penicillin
3 ❏ benzathine penicillin
A ❏ 1
B ❏ 2
C ❏ 3

D ❑ 1 and 2
E ❑ 1, 2 and 3

Q76 The dip tube in an aerosol container is made from:

A ❑ glass
B ❑ stainless steel
C ❑ polypropylene
D ❑ aluminium
E ❑ tin

Q77 The flash point of an aerosol product is determined using:

A ❑ pressure gauze
B ❑ hydrometer
C ❑ tag open-cup apparatus
D ❑ Freon dispenser
E ❑ Karl Fischer apparatus

Q78 Opening and closing of the valve in an aerosol container is regulated by:

A ❑ actuator
B ❑ gasket
C ❑ ferrule
D ❑ valve seal
E ❑ valve stem

Q79 Glycols such as polyethylene glycols are integral components of which of following type of foam?

A ❑ aqueous stable
B ❑ non-aqueous stable
C ❑ quick-breaking
D ❑ thermal
E ❑ perfume

Q80 Which of the following devices relies on patient inspiration to withdraw drug from the device?

A ❑ dry powder inhaler
B ❑ metered-dose inhaler
C ❑ jet nebuliser
D ❑ ultrasonic nebuliser
E ❑ pneumatic nebuliser

Q81   The chemical name of propellant 114 is:

A   ❏ dichlorodifluoromethane
B   ❏ trichloromonofluoromethane
C   ❏ difluoroethane
D   ❏ heptafluoropropane
E   ❏ dichlorotetrafluoroethane

Q82   A breath-actuated metered-dose inhaler is:

A   ❏ Autohaler™
B   ❏ Rotahaler™
C   ❏ Spinhaler™
D   ❏ Turbohaler™
E   ❏ Dryhaler™

Q83   An example of a compressed gas propellant is:

A   ❏ butane
B   ❏ isobutane
C   ❏ nitrous oxide
D   ❏ dimethyl ether
E   ❏ propane

Q84   Aerosol valve discharge rate is expressed as:

A   ❏ g/mL
B   ❏ g/s
C   ❏ mL/s
D   ❏ mL/mm
E   ❏ mL/g

Q85   The propellants being phased out under the terms of the Montreal Protocol include:

A   ❏ chlorofluorocarbons
B   ❏ hydrocarbons
C   ❏ hydrocarbon ethers
D   ❏ compressed gases
E   ❏ nitrous oxide

Q86   An aerosol preparation with a high proportion of propellant 12 requires filling in the following types of container:

1   ❏ aluminium
2   ❏ plastic

3 ❑ coated glass

4 ❑ stainless steel

A ❑ 1

B ❑ 1 and 3

C ❑ 3 and 4

D ❑ 1 and 4

E ❑ 4

Q87 The particle size of pharmaceutical aerosols is measured by:

1 ❑ cascade impactor

2 ❑ gas chromatography

3 ❑ infrared spectroscopy

A ❑ 1

B ❑ 2

C ❑ 1 and 2

D ❑ 2 and 3

E ❑ 1, 2 and 3

Questions 88–90 involve the following case:

A pharmaceutical company is involved in the production of therapeutically active ingredients in a pressurised system or aerosol, defined as a system that depends on the power of a propellant to expel the contents from the container. The products prepared are in the form of suspended powder in water-based aerosols, intended for oral and inhalational use only.

Q88 The aerosol system formed is:

A ❑ a two-phase system

B ❑ a three-phase system

C ❑ a macrophase

D ❑ a homogeneous phase

E ❑ a double phase

Q89 A propellant that can be used is:

A ❑ hydrocarbons

B ❑ hydrocarbon ether

C ❑ fluorinated hydrocarbons

D ❑ compressed gases

E ❑ nitrous oxide

Q90 The propellant can be filled by:

A ❏ cold filling technique
B ❏ dosator method
C ❏ compressed gas filling technique
D ❏ hot fill method
E ❏ none of the above

Q91 Copyright is applicable to original works of all of the following categories except:

A ❏ literary works
B ❏ artistic works
C ❏ musical works
D ❏ court and tribunal judgments
E ❏ films

Q92 Which one of the following is not a legal requirement for getting a utility patent?

A ❏ usefulness
B ❏ novelty
C ❏ nationality of inventor or applicant
D ❏ non-obviousness
E ❏ fitting in to a statuary class

Q93 PCT is the:

A ❏ Patent Coordination Team
B ❏ Patent Cooperation Team
C ❏ Patent Cooperation Treaty
D ❏ Patent Commission Treaty
E ❏ Patent Commission Team

Q94 GATT is the:

A ❏ General Article on Trade and Tariff
B ❏ General Agreement on Trade and Transport
C ❏ General Article on Tariffs and Trade
D ❏ General Agreement on Traffics and Trade
E ❏ General Agreement on Tariffs and Trade

Q95 TRIPS is:

A ❏ Transport-Related Aspects of Intellectual Property Rights
B ❏ Tariffs-Related Aspects of Intellectual Property Rights

C ❏ Trade-Related Agreement on Intellectual Property Rights
D ❏ Trade-Related Aspects of Intellectual Property Rights
E ❏ Tariffs-Related Agreement on Intellectual Property Rights

Q96 WIPO is:

A ❏ World Intellectual People Organization
B ❏ World Intellectual Property Organization
C ❏ World Intelligence Police Organization
D ❏ World Intellectual Programme Ordinance
E ❏ World Intelligence Possession Organization

Q97 Which of the following is (are) a type(s) of patent?

1 ❏ utility
2 ❏ design
3 ❏ plant
A ❏ 1
B ❏ 1 and 2
C ❏ 1 and 3
D ❏ 2 and 3
E ❏ 1, 2 and 3

Q98 When an invention performs substantially the same function in substantially the same manner and obtains the same result as the patented invention, this type of patent infringement is called:

1 ❏ literal infringement
2 ❏ equivalent infringement
3 ❏ inducing infringement
4 ❏ doctrine of equivalents (DoE)
A ❏ 2
B ❏ 1 and 2
C ❏ 1 and 3
D ❏ 1 and 4
E ❏ 2 and 4

Questions 99 and 100 involve the following case:

A scientist gets a patent on their invention and gets a registered trademark on their invented product in their own country.

Q99   How long does a patent protection last?

A   ❑ 1 year
B   ❑ 5 years
C   ❑ 10 years
D   ❑ 20 years
E   ❑ for ever

Q100   How long does a trademark registration last?

A   ❑ 1 year
B   ❑ 5 years
C   ❑ 10 years
D   ❑ 20 years
E   ❑ for ever

Answers

A1   B

Gravitational sedimentation is the primary deposition mechanism for particles in the size range 0.5–3 µm, in the small airways and alveoli, for particles that have escaped deposition by impaction. Brownian diffusion is the predominant mechanism for particles smaller than 0.5 µm. Interception of particles with extreme shapes and electrostatic attraction with appropriately charged particles are less predominant mechanisms of deposition.

A2   D

Particles sized approximately 0.5 µm are inefficiently deposited, being too large for effective deposition by Brownian diffusion and too small for effective impaction or sedimentation, and are often immediately exhaled. All particles are affected by hygroscopic growth, particularly if they are water-soluble.

A3   C

The most important physical property of an aerosol for inhalation is the particle size. Particles with a diameter greater than 5 µm are likely to deposit via inertial impaction in the upper airways (including the mouth). Increasing the hydrophilicity of the active ingredient will result in increased hygroscopic growth and therefore increased particle size. Decreasing the particle size will enable deposition via sedimentation, with less active ingredient deposited in the mouth due to impaction. The particle size of the excipients will not affect deposition of the active drug.

A4   E

When a hydrophilic material enters the respiratory tract, the increase in relative humidity leads to condensation of water on the particle surface. Since the vapour pressure of the solution formed on the particle surface is lower than that of pure solvent, water will continue to condense and the particle size will increase until equilibrium between vapour pressures is reached. The larger particle will be deposited higher in the respiratory tract.

A5   B

Sodium cromoglycate is poorly absorbed orally. The lung is useful for delivering drugs with systemic activity due to the abundance of capillaries and large surface area (70–$80\,m^2$ in adult males). The lung itself has some metabolic activity.

A6   A

HFAs are poor solvents for the surfactants commonly used in MDIs, have lower boiling points than CFC-11 and CFC-114, and contribute to global warming. Hydrocarbon propellants such as propane and butane are not used in inhalation products due to their toxicity and flammability. Ethanol is included in formulations containing HFAs to allow dissolution of surfactants; however, due to its low volatility, it may increase the droplet size of the emitted aerosols.

A7   A

MDIs are hermetically sealed and the propellant vapour is inert, therefore protecting ingredients from oxidative degradation and microbial contamination. An increase in humidity affects DPIs, since the powders may clump. The propellants used in MDIs are poor solvents for most active ingredients. MDIs are inefficient at drug delivery, since propellant droplets exit at high velocity, resulting in the active drug being lost due to impaction.

A8   A

Whereas ultrasonic nebulisers increase solution temperature, jet nebulisers may cause a decrease of 10–15 °C and patients may experience broncho-constriction when inhaling colder air. Preservatives may cause bronchospasm. Unit dose formulations are sterile and preservative-free. Approximately 1 mL of nebuliser fluid remains as 'residual' or 'dead' volume and is unable to be nebulised in jet nebulisers; therefore, small-volume nebuliser fluids may be diluted to a larger volume by adding sterile saline, thus enabling most of the therapeutic dose to be delivered.

A9   C

The rate of gas flow is the major determinant of the aerosol droplet size and the rate of drug delivery for jet nebulisers – as the flow rate increase, the MMAD decreases and the rate of drug delivery increases. Nebulisers have been developed (LC nebuliser™) whereby the patient's own breath boosts nebuliser performance, with aerosol production matching the patient's tidal volume, thus minimising wastage, since the inspiratory phase of breathing constitutes only one-third of the breathing cycle. Nebulisers deliver larger volumes of drug compared with DPIs and MDIs.

A10   E

Piezoelectric crystals are used in both ultrasonic and vibrating-mesh nebulisers. The advantage of nebulisers compared with MDIs and DPIs is that the drug is inhaled during normal tidal breathing and does not require inhalation/actuation coordination. Ultrasonic nebulisers may increase the solution temperature by up to 10–15 °C, whereas jet nebulisers result in a decrease in solution temperature and decreased drug solubility. Vibrating-mesh nebulisers are reported to have a negligible effect on fluid temperature.

A11   D

Suspensions are poorly delivered from ultrasonic nebulisers. Antioxidants, particularly sulfites, may cause bronchospasm. Solutions with high hydrogen ion concentrations may cause bronchoconstriction. Due to elevated temperatures produced by ultrasonic nebulisers, peptides and proteins are denatured. The cooling effect of jet nebulisers may result in increased liquid surface tension and viscosity.

A12   E

Fentanyl is a suitable candidate for transdermal delivery because of its limited solubility in water, it is un-ionised, it has a suitable molecular weight (range 100–800) and it is potent. The therapeutic index is not a factor determining the suitability of drugs for transdermal delivery.

A13   E

Penetration of the stratum corneum is determined by the drug concentration (the higher the concentration, the greater the penetration) and the partition coefficient of the drug between the tissue and the vehicle. Penetration of the stratum corneum needs to occur in order for the drug to act systemically, and it proceeds by a passive process. Greater penetration occurs when the skin is damaged or abraded. The vehicle has a role to play in drug penetration, since, unlike many other vehicles of pharmaceutical products, it remains in contact with the delivery site (skin).

A14   C

The larger the drug partition coefficient between the vehicle and the skin, the greater the flux of the drug across the skin. Drug flux involving bidirectional movement of the drug substance at the membrane or surface is possible. The larger the drug diffusion coefficient, the greater the drug flux into the skin; the greater the thickness of the stratum corneum, the less the drug flux into the skin; and the greater the difference in drug concentration between the vehicle and the tissue, the greater the drug flux.

A15   A

Increased hydration would facilitate transport of fentanyl across the skin, whereas increased molecular size would not facilitate the transport of the drug through the skin – range 100–800, with optimum around 400. Other factors such as increased temperature, pH and drug concentration all facilitate the transport of drugs across the skin.

A16   B

Transport of water-soluble drugs is transcellular and not through the endogenous lipid. When the drug is more soluble in the stratum corneum, it presents as a higher concentration in the first layer of the skin. If depletion of the contact layer of the vehicle occurs, then the nature of the formulation will dictate how readily these are replenished by diffusion and therefore will dictate the rate of absorption.

A17   D

In this system, drug release is determined by the rate-controlling membrane, which contains the drug reservoir in a liquid or gel form. Drug matrix (polymeric material) is between the backing and frontal layers; this matrix controls the release of the drug in a monolithic system and not a membrane-controlled system – in monolithic systems, the Higuchi square root of time law is usually obeyed.

A18   E

An advantage of transdermal delivery is that there can be rapid identification of the drug, especially if the patient is unconscious. The fact that this route avoids gastrointestinal drug absorption difficulties, permits extended therapy and thereby improves compliance, and can be terminated rapidly are all advantages of this delivery route.

A19   E

Combinations of penetration enhancers such as iontophoresis and chemicals may result in synergy and limit the side-effects associated with the system. Iontophoresis is an electrochemical method using an applied electrical current and as such enhances the transport of charged or ionised drugs. The process involves electrostatic repulsion at the electrode, with negative ions delivered by the cathode, while positive ions are delivered by the anode.

A20   A

Since fentanyl citrate is the only ionised (charged) drug, it is suitable for iontophoretic delivery. Since morphine, indometacin, diclofenac and ketoprofen are all un-ionised, they do not require iontophoresis for transdermal delivery.

A21   B

The hydrophilic gelling agent carbomer 934 controls the release of the drug. Estradiol is the active drug and triethanolamine modifies the pH, because the pH needs to be basic in order for the carbomer to gel. Propylene glycol and ethanol have co-solvent and preservative properties.

A22   B

Suppositories may exert local and systemic action. Local action may involve the relief of constipation or the pain, irritation and inflammation associated with haemorrhoids or anorectal conditions. Antihaemorrhoidal suppositories contain local anaesthetics, vasoconstrictors, analgesic, soothing emollients and protective agents.

A23   E

Disadvantages associated with the use of suppositories include strong feelings of aversion in certain countries. More rationally, their effects are slow, with incomplete absorption and considerable inter- and intra-patient variations.

A24   A

Cocoa butter melts easily. Because of its immiscibility with body fluids, it fails to release fat-soluble drugs. To maximise bioavailability for systemic drug action, incorporate the ionised rather than the un-ionised form of the drug.

A25   C

Physicochemical factors of the drug and the suppository base that are important include the lipid/water solubility (lipophilic drugs are not released from fatty bases, while water-soluble bases release both lipophilic and hydrophilic drugs), particle size (the smaller the particle size, the more readily the dissolution and subsequent absorption), and the nature of the suppository base (the base must be capable of melting, softening or dissolving in order to release its drug components for absorption).

A26   D

Synthetic fatty bases on melting become more fluid than theobroma oil, and thus sedimentation of drugs in this base is greater. Because this may result in uneven distribution of the drug in the suppository, thickeners may be added to ensure that the drug is evenly distributed throughout the suppository.

A27   D

Suppository bases should melt at body temperature, dissolve or disperse in body fluids, retain their shape and be stable at room temperature. They should also be non-toxic and non-irritant to the mucous membranes of the rectum. Bases are also required to be compatible with any added drug and to release that drug from the base.

A28   A

Macrogols are commonly used in suppository bases as two or more grades can be used to achieve a suitable melting point and physical properties. Disadvantages are that they are hygroscopic, have good solvent properties (which results in retention of the drug in the base in the body), sometimes fracture on storage, and may promote crystal growth of the drug within the suppository.

A29   D

Bioavailability of drugs administered by the vaginal route is improved by bypassing the first-pass effect. The vaginal wall is well suited to absorption as it has a vast network of blood vessels. However, few drugs are administered in this way – mainly oestrogens as vaginal creams and gels. These preparations do come into contact with tissues that are prone to infection and therefore need to be free of offending organisms; therefore, vaginal gels are required to be preserved with antimicrobial agents.

A30   B

Some vaginal inserts are capsules of gelatin containing a medicament to be released intravaginally. These inserts are made of gelatin to which glycerin or a polyhydric alcohol such as sorbitol has been added to render the gelatin elastic- or plastic-like. They do, however, contain more moisture than hard gelatin capsules and are therefore required to be preserved by methyl or propyl paraben. They may be used to encapsulate liquids, suspensions, pasty materials and dry powders, and they are elegant and easily applied by the patient.

A31   E

Vaginal inserts/tablets may be used more widely than pessaries due to the fact that they are easier to manufacture, stable and less messy. Although they are prepared by standard compression using standard excipients, additional excipients need to be added to facilitate disintegration due to the low moisture content of the vagina.

A32   A

Proteins are delicate, large molecules with many functional groups. Because the three-dimensional structure is stabilised by weak physical bonds, it can readily and irreversibly be altered, with this alteration affecting the interaction at the receptor, affecting the pharmacokinetic parameters and potentially making the protein immunogenic. Transporter molecules are often required to facilitate epithelial penetration, which otherwise is extremely slow.

A33   E

Physical instability can be caused by elevated temperatures (causing denaturation), low temperatures (causing destabilisation) and shaking (resulting in the protein unfolding and aggregation occurring).

A34   D

Reducing sugars such as lactose should not be used as excipients in protein formulations as they can react with free primary amino groups of the protein molecule via a Maillard reaction to form brown-coloured reaction products. The role of the other excipients is as follows: amino acids – enhance solubility; detergents – antiadsorbents; phosphate – buffer; ascorbic acid – antioxidant.

A35   A

Although the proteins degrade rapidly, they may not be stabilised by refrigeration, because low temperatures, unlike for most other drugs, cause destabilisation. It is preferred that pharmaceutical proteins are stored in the dry form (preferentially by freeze drying with a choice of excipients such as lyoprotectants being extremely important) and reconstituted. This allows the preferred shelf life of a minimum of 2 years to be achieved.

A36   E

The naturally occurring protein is the L-form. Isomerisation from the L- to the D-form will change the structure of the protein. Deamidation, a common degradation reaction in water, is dependent on pH and the neighbouring amino acids. Oxidation is catalysed by traces of transition metals, and in an oxidative environment free cysteine units may form disulfide bridges. Under neutral conditions, most of the peptide bonds between amino acids are stable.

A37   A

Lyoprotectants are non-reducing sugars. Their role is to stabilise the product as the removal of water may irreversibly affect the protein structure.

A38   B

Although the nasal, rectal, buccal and transdermal routes are easily accessible, the pulmonary route is the only alternative route that does not result in low bioavailability of proteins.

A39   C

Since it is important to guarantee the integrity of the protein, a set of pharmacological, immunological, spectroscopic, electrophoretic and chromatographic approaches are needed to characterise the protein as closely as possible. ELISA is an example of an immunological test that is performed.

A40   D

Protein-bound drugs are incapable of penetrating the corneal epithelium because of the size of the protein–drug complex. Tears may contain enzymes such as lysozyme, and these are capable of metabolic degradation of drug substances. Ophthalmic suspensions, gels and ointments do not mix readily with the lacrimal fluids, and this causes them to remain longer in the cul-de-sac, thus enhancing drug activity.

A41   C

Since the normal volume of tear fluid in the cul-de-sac of the human eye is 7–8 µL, much of a single drop of an ophthalmic solution (50 µL) is lost on administration. Ophthalmic solutions are not well retained on the eye, and this results in frequent dosing being required. However, this can be addressed by formulation of suspensions, gels and ointments.

A42   C

Filtration is a method using for sterilisation of ophthalmic preparations, especially in a practice setting. Dry heat is not suitable because of the thermolabile nature of the active ingredient, while the complexity of the other methods precludes their use in a practice setting.

A43   B

pH adjustment and buffering enhances the aqueous (not the lipid) solubility of a drug and as such provides formulation stability, provides comfort to the eye, enhances bioavailability and maximises preservative efficacy.

A44   D

When preparing a product that is a suspension for ocular delivery, the suspended particles should not aggregate and need to redisperse easily on shaking in order to deliver an accurate dose.

A45   D

Convenience, a good area for absorbing drugs, sufficient blood supply and a nasal cavity lined with mucous membrane are all advantages of nasal administration of drugs. However, the presence of cilia facilitates the movement of mucus, resulting in mucociliary clearance, which is a non-specific defence action and thus can also act as a defence against drugs.

A46   E

Since the mucus in the nasal cavity is a diffusional barrier to drug absorption, it is important that formulation of drugs overcomes this barrier such that the drug remains in contact long enough to be released and absorbed. However, it is also important that the interruption of the mucociliary clearance by the drug or excipients is minimal or temporary. Nasal mucosa possesses enzymatic activity offering protection against exogenous chemicals. Nasal first-pass metabolism is a significant factor in the absorption of drugs.

A47   A

Phenylephrine hydrochloride (active drug), sodium metabisulfite (antioxidant), sodium chloride (adjusting for isotonicity) and propylene glycol (co-solvent) are components of the formulation. Chlorobutanol (chlorbutol) is the preservative in the nasal spray.

A48   C

Molecular weight and size, pH and lipid partition coefficient, solubility, powder morphology and particle size of the drug are all physicochemical properties of the drug that affect nasal absorption. The drug class is not a physicochemical factor and thus does not affect nasal absorption.

A49   E

Absorption enhancers increase the rate at which drugs pass through the nasal mucosa, especially for those peptides and proteins where size leads to poor bioavailability. Surfactants cause mucosal damage and are therefore not suitable as absorption enhancers, whereas bile salts enhance activity but are less damaging. For cyclodextrins, accommodation of hydrophobic drugs entirely or partially within the cavity of the cyclodextrin molecule increases solubility and thus bioavailability.

A50   C

Metered-dose pump systems may be formulated as solutions, suspensions or emulsions, delivering a predetermined volume of 25–200 μL. Deposition of the drug occurs over a large surface area and the particle size is important. The volume delivered is affected by the size of the nasal cavity.

A51   B

Buccal tablets are designed for absorption through the oral mucosa but are designed to erode slowly. They allow the oral absorption of drugs that are destroyed and poorly absorbed in the gastrointestinal tract. Lozenges and troches are other solid dosage forms suitable for buccal delivery.

A52   E

Parenteral formulations offer the distinct advantages of rapid onset of action, reliability and predictability, and avoidance of the gastrointestinal tract.

A53   E

Despite the unique advantages offered by parenteral preparations, there are limitations of cost, needle fear, pain, and difficulty in correcting errors of dosing.

A54   D

Intramuscular injections do not offer 100% bioavailability similar to formulations administered by the intravenous route.

A55 E

Parenteral formulations exist in the form of suspensions, solutions and emulsions, although solutions are the most common form.

A56 B

Calcium ions adversely affect the stability of intravenous lipid emulsions.

A57 C

Intravenous administration provides the quickest route for administration in most cases. The highest plasma concentration is achieved within 4 min of intravenous administration, but this can be in the order of several hours with oral administration.

A58 A

LVPs are not buffered, because of the large quantity of buffer salts required and the potential for toxicity.

A59 D

Dissolved oxygen can be a major cause of drug degradation in parenteral products and attempts are made to minimise its levels.

A60 B

Electrolytes such as sodium chloride are used to adjust pH and tonicity, which are critical for parenteral preparations.

A61 B

The subcutaneous route provides a slower onset of action than the intramuscular or intravenous routes. The maximum volume that can be injected by this route is approximately 1 mL, e.g. vaccines, insulin and adrenaline. Typically, the injection site is rotated for daily or frequent administration (e.g. in patients with insulin-dependent diabetes).

A62 E

Although the onset is very rapid, the duration of action of an intravenous preparation is a function of the drug, its administration and its pharmacokinetic profile.

A63 B

Muscle tissue is more acidic compared with most biological fluids, meaning that a lower pH will support un-ionisation of weakly acidic drugs.

A64 E

Chlorinated solvents are not used in parenteral preparations. There are strict regulatory limits on residual solvents in pharmaceutical products.

A65 B

Sterile filtration is a universally accepted method of preparing parenteral products in addition to autoclaving.

A66 B

The usual size for lipid globules is 0.3–0.4 μm.

A67 E

Trace elements tend to destabilise complex lipid emulsions.

A68 E

Any factor that can increase levels of oxygen will tend to destabilise oxidisable compounds.

A69 C

A small cannula size will help in minimising tissue damage and the resulting thrombophlebitis.

A70 E

Particulate contamination can also result because of a number of additional sources, including impurities, processing equipment and filtration devices.

A71 E

A major limitation of parenteral formulations is the difficulty of fixing errors that may happen because of miscalculations or pumping-system errors leading to overloading. Although allergy can occur with any route of administration, thrombophlebitis is related to the use of parenteral preparations.

A72   C

Intrathecal is a specialised route of administration requiring close monitoring and is used for very few drugs.

A73   E

A number of drugs and dosage-form-related factors, as listed in the three options, can influence parenteral drug absorption, similar to other routes of administration.

A74   C

The intravenous route does not provide the option of controlled drug release because of the route of administration. The intramuscular route is the most commonly used parenteral route for controlled drug release.

A75   E

Ester, salt or other modification may be employed to increase stability, alter drug solubility, enhance depot action, ease formulation difficulties, and possibly decrease pain on injection. In these examples, the solubility of the drug is modified to provide controlled release.

A76   C

A dip tube extends from the bottom of the aerosol container and allows for the flow of product when the valve is opened. Dip tubes are made from polypropylene or polyethylene, although polypropylene tubes are usually more rigid. Polypropylene is a thermoplastic polymer resistant to many chemical solvents, bases and acids.

A77   C

The flash point of an aerosol product is the lowest temperature at which the volatile propellant can vaporise to form an ignitable mixture in air. Measuring an aerosol's flash point requires an ignition source. There are two basic types of apparatus for flash point measurement – tag open-cup apparatus and closed-cup apparatus. In tag open-cup apparatus, the sample is contained in an open cup which is heated, and at intervals a flame is brought over the surface.

A78   A

Aerosol containers have three major parts: the can, the valve and the actuator or button. An actuator is a mechanical device for moving or controlling a mechanism or system. The actuator is depressed by the user to open the valve. The shape and size of the nozzle in the actuator control the spread of the aerosol spray. The nozzle also aids in producing the required type of product discharge.

A79   B

Non-aqueous solvents such as glycols and polyethylene glycol form stable foams and are the integral components of non-aqueous stable foams. Non-aqueous stable foams are generally formulated according to the following:

Glycol	91.0–92.5% w/w
Emulsifying agent	4.0% w/w
Hydrocarbon propellant	3.5–5.0% w/w

The emulsifying agents found to be most effective were from the class of glycol esters, e.g. propylene glycol monostearate.

A80   A

A dry powder inhaler (DPI) is a device that delivers medication to the lungs in the form of a dry powder. DPIs rely on the force of patient inhalation to entrain powder from the device and subsequently break up the powder into aerosol particles. The medication is commonly held in a capsule for manual loading. Once loaded or actuated, the operator puts the mouthpiece of the inhaler into their mouth and takes a deep inhalation, holding their breath for 5–10 s. Thus, most DPIs have a minimum inspiratory effort that is needed for proper use; for this reason, DPIs are normally used only in older children and adults.

A81   E

Propellants are designated by three digits (000). The first digit represents one less than the number of carbon atoms $(C-1)$ in the compound. The second digit represents one more than the number of hydrogen atoms $(H+1)$ in the compound. The third digit represents the number of fluorine atoms in the compound. The number of chlorine atoms in the compound can be found by subtracting the sum of the fluorine atoms and hydrogen atoms from the total number of atoms that can be added to saturate the carbon chain. So, for propellant 114, the number of carbon atoms is 2, the number of hydrogen atoms is 0, the number of fluorine atoms is 4, and the number of chlorine atoms is 2; therefore, propellant 114 is dichlorotetrafluoroethane $(C_2Cl_2F_4)$.

A82   A

Autohaler^{TM}, a product of 3M Pharmaceuticals, is a breath-actuated MDI in which the patient inhalation triggers the dose. All the other devices mentioned are dry powder inhalers.

A83   C

Nitrous oxide, nitrogen and carbon dioxide are compressed gases used as aerosol propellants. Compressed gases have little expansion power. Compressed gases are used in topical products such as hair preparations, ointments and dental creams, since they lack a chilling effect.

A84   B

Aerosol valve discharge rate is determined by taking an aerosol product of known weight and discharging the contents for a given period of time using standard apparatus. By reweighing the container after the time limit has expired, the change in weight per time dispensed is the discharge rate, which can then be expressed as grams per second (g/s).

A85   A

The chlorofluorocarbons are phased out under the terms of the Montreal Protocol due to their ozone-depleting potential.

A86   D

Propellant 12, or dichlorodifluoromethane, has a high vapour pressure which further increases with concentration. Aerosol preparations containing a high proportion of propellant 12 require the contents to be packaged in metal containers of aluminium or stainless-steel that must withstand the high pressure.

A87   A

A cascade impactor is a multistage sampling device for determining the size distribution of a particulate aerosol. A cascade impactor uses the principle of inertial separation to size-segregate particle samples from a particle-laden gas stream. The aerosol flows into the impactor, where it impinges upon a sequence of solid discs (stage). Each disc is contained within a flow chamber. Each chamber is connected in a vertical arrangement to the previous and next chamber in the sequence. Larger particles impact on the first disc and are captured. The sampling velocity increases for each successive chamber/disc, so that successively smaller particles are collected. The weight of each size fraction is then determined gravimetrically.

A88   B

Water-based suspended powder aerosols form three-phase aerosol systems since propellant and water are not miscible. The existence of propellant phase, water phase and vapour phase (three phases) is a characteristic of suspended powder aerosols.

A89   C

Aerosols for oral or inhalational use have been exempted from the Food and Drug Administration (FDA) ban. The fluorinated hydrocarbons, namely propellants 12, 12/114 and in some cases 12/11, are used.

A90   A

Fluorinated hydrocarbons, although heavier than air, do not form explosive or flammable mixtures and are filled using cold filling apparatus.

A91   D

The term 'copyright' refers to the legal right to exclude others, for a limited time, from copying, selling, performing, displaying or making derivative versions of a work of authorship. Copyright can exist in original works such as literary works, dramatic works, artistic works, musical works, sound recordings, films, broadcasts, cable programmes and typographical arrangements of published editions. However, copyright protection does not apply to certain government works such as parliamentary bills, Acts of Parliament, regulations, by-laws, parliamentary debates, select committee reports, court and tribunal judgments, reports of royal commissions, commissions of inquiry, ministerial inquiries and statutory inquiries.

A92   C

To be patented, the invention must fit into one of the statutory classifications (i.e. processes, machines, articles of manufacture, compositions or 'new uses' of one of the first four), and it must be useful, novel and non-obvious. The applicant must qualify as a true inventor of the invention. Personal qualities such as age, sex, citizenship, country of residence, physical disabilities, health, mental competence, incarceration, nationality, religion, race and creed are irrelevant.

A93   C

The Patent Cooperation Treaty (PCT) is an international patent law treaty that provides a unified procedure for filing patent applications to protect inventions in each of its contracting states. A patent application filed under the PCT is called an international application or PCT application. This treaty was concluded in 1970.

A94   E

The General Agreement on Tariffs and Trade (GATT) is a multilateral agreement regulating trade among about 150 countries. According to its preamble, the purpose of the GATT is the 'substantial reduction of tariffs and other trade barriers and the elimination of preferences, on a reciprocal and mutually advantageous basis'. GATT was formed in 1947 and lasted until 1994; it was replaced by the World Trade Organization (WTO) in 1995.

A95   D

The Trade Related Aspects of Intellectual Property Rights (TRIPS) Agreement is an international agreement, negotiated in 1994 and administered by the World Trade Organization (WTO), that sets down minimum standards for many forms of intellectual property (IP) regulation. It introduced IP law into the international trading system for the first time and remains the most comprehensive international agreement on IP to date.

A96   B

The World Intellectual Property Organization (WIPO) is one of the 16 specialised agencies of the United Nations. WIPO was created in 1967 'to encourage creative activity, to promote the protection of intellectual property throughout the world'.

A97 E

There are three types of patent – utility patents, design patents and plant patents. Utility patent is the main type of patent, which covers inventions that function in a unique manner to produce a utilitarian result. Design patent covers the unique, ornamental, or visible shape or design of a non-natural object. Plant patent covers plants that can be reproduced through the use of grafts and cuttings (asexual reproduction).

A98 E

Doctrine of equivalents (DoE), also known as equivalent infringement, is a form of direct patent infringement that occurs when an invention performs substantially the same function in substantially the same manner and obtains the same result as the patented invention. A court analyses each element of the patented invention separately.

A99 D

A patent protection lasts for 20 years from the date that the intellectual property office (IPO) of the country receives a complete application, provided that the renewal fees are paid at the specified time periods of the patent's existence.

A100 C

A trademark registration lasts for 10 years from the date that the intellectual property office (IPO) of the country receives the application. The registration is thereafter renewable for further periods of 10 years.

Test 6

Miscellaneous topics

Sanjay Garg, Therese Kairuz and Roop K Khar

Pharmaceutical microbiology

Regulatory affairs

Good manufacturing practices (GMP)

Good clinical practices (GCP)

Extemporaneous compounding

Quality assurance (QA) and quality control (QC)

Good laboratory practices (GLP)

Packaging

Introduction

This section includes topics that are covered to varying degrees in pharmacy programmes around the world. Microbiology, extemporaneous compounding, regulatory affairs and packaging are integral components of drug development and the manufacturing process. A number of pharmaceutical preparations such as parenteral products, ophthalmic formulations, dialysis solutions and implants are required to be sterile. Pharmacopoeias generally prescribe the methods of sterility testing and preservative effectiveness. The methods of sterilisation used in pharmaceutical industry include dry heat, autoclaving, filtration and gamma radiation. The tests for particulate matter, clarity of solution and pyrogens are additional tests that complement microbial testing. The microbiological methods are also used to determine the potency of antimicrobial drugs and preservatives, and as a quality control (QC) tool.

Extemporaneous compounding is commonly used in some countries, while in other countries it has been largely replaced by commercially available pharmaceutical products. In the absence of suitable formulations, pharmacists are often asked to modify the available products to suit the individual patient's requirements. For example, tablets for adults can be crushed and suspended in syrup base for paediatric patients. These formulations usually lack standardised recipes and stability information, and are often characterised by a short shelf life.

Quality, safety and efficacy are three pillars of the healthcare industry and the responsibility of everyone involved with medicines and patients. In pharmacy, quality is defined as compliance to requirements or specifications. Medicines are unique because their quality can not be judged by patients and pharmacists in most cases. Any compromise on quality can be fatal to patients. Pharmaceutical regulatory affairs are an integral part of the pharmaceutical industry and offer numerous job opportunities to pharmacists. Regulatory agencies around the world, such as the US Food and Drug Administration (FDA) and the Therapeutic Goods Administration (TGA), Australia, employ a number of pharmacists in drug registration and inspection units. The training on good practice (GXP), i.e. good laboratory practices (GLP), good manufacturing practices (GMP) and good clinical practices (GCP), and their basic components, i.e. standard operating procedures and documentation, empower pharmacists to ensure the quality, safety and efficacy of medicines. These systems also extend to pharmacies and the hospital environment. The understanding is also very useful to pharmacists working with new drug discovery and development programmes.

Quality assurance (QA) is a wide-ranging concept and refers to the sum total of organised arrangements made to ensure the quality of medicines. It encompasses GMP as well as QC. Quality control involves assessing the quality of active ingredients, excipients, packaging components, processing conditions and finished products. Good clinical practices deal with the design, conducting, monitoring, auditing and recording of clinical studies.

Developments over the past few years, such as process analytical technology, quality by design and quality risk management, are making the field of pharmaceutical regulatory affairs an interesting one to learn and practise.

Questions

Q1   For the validation of dry heat sterilisation, spores of which of the following are used as biological indicators?

A   ❏ *B. subtilis*
B   ❏ *B. stearothermophilus*
C   ❏ *B. pumilus*
D   ❏ *C. welchii*
E   ❏ *B. tuberculosis*

Q2 The cell wall of a Gram-positive bacteria typically:

A ❑ consists predominantly of glycoproteins
B ❑ lacks peptidoglycans
C ❑ has peptidoglycan-containing muramic acid and D-amino acid
D ❑ has repeating units of arabinogalactan and ungeolates
E ❑ consists predominantly of polysaccharides

Q3 Which of the following immunising agents is administered orally?

A ❑ rabies vaccine
B ❑ polio vaccine
C ❑ tetanus toxoid
D ❑ mumps virus vaccine
E ❑ DPT vaccine

Q4 Which of the following is the immunoglobulin (Ig) responsible for the primary immune response?

A ❑ IgA
B ❑ IgB
C ❑ IgD
D ❑ IgE
E ❑ IgM

Q5 The percentage efficiency (%) of high-efficiency particulate air (HEPA) filters in removing particles $\geq 0.3\,\mu m$ is:

A ❑ 99.97
B ❑ 95.0
C ❑ 90.0
D ❑ 98.3
E ❑ 50.0

Q6 The vector for dengue is:

A ❑ mansonia
B ❑ anopheles
C ❑ culex
D ❑ aedes
E ❑ tsetse fly

Q7 BCG is a:

A ❑ live viral vaccine
B ❑ live bacterial vaccine
C ❑ toxoid of bacterial origin
D ❑ attenuated viral vaccine
E ❑ killed viral vaccine

Q8 The time interval required, at a specified constant temperature, to reduce the microbial population by 90% is the:

A ❑ D value
B ❑ Z value
C ❑ T value
D ❑ F value
E ❑ N value

Q9 Which of the following gases is used for medical product sterilisation?

A ❑ argon
B ❑ xenon
C ❑ ethylene oxide
D ❑ halothane
E ❑ isopropane

Q10 Typhus fever is caused by

A ❑ *Salmonella typhi*
B ❑ *Bordetella pertussis*
C ❑ *Yersinia pestis*
D ❑ *Chlamydia trachomatis*
E ❑ *Rickettsia prowazekii*

Q11 The route of transmission for *Coxiella burnetii*, a causative organism of Q fever is:

A ❑ faecal
B ❑ insect vector
C ❑ topical
D ❑ inhalational
E ❑ peroral

Q12 The herpes virus has:

A ❑ double-stranded DNA
B ❑ single-stranded DNA

C   ❑ double-stranded RNA
D   ❑ single-stranded RNA
E   ❑ single-stranded DNA and RNA

Q13   The triple vaccine is used for:

A   ❑ diphtheria, tetanus, polio
B   ❑ diphtheria, tetanus, smallpox
C   ❑ diphtheria, tetanus, pertussis
D   ❑ diphtheria, pertussis, polio
E   ❑ tetanus, pertussis, polio

Q14   The Widal test is positive for which of the following infections?

A   ❑ pseudomonas
B   ❑ shigella
C   ❑ trachoma
D   ❑ salmonella
E   ❑ rickettsia

Q15   Interferons are chemicals that:

A   ❑ act on adjacent cells and render them refractory to viral damage
B   ❑ interfere with the multiplication of viruses
C   ❑ kill viruses
D   ❑ kill bacteria
E   ❑ kill fungi

Q16   Commercial production of citric acid is carried out by the microbial culture of:

A   ❑ *Fusarium moniliforme*
B   ❑ *Rhizopus nigricans*
C   ❑ *Candida utilis*
D   ❑ *Saccharomyces cerevisiae*
E   ❑ *Aspergillus niger*

Q17   Parenteral formulation of the antibiotic 'penicillin' is manufactured in a:

A   ❑ class 1 room facility
B   ❑ class 10 room facility
C   ❑ class 100 room facility
D   ❑ class 1000 room facility
E   ❑ class 10 000 room facility

Q18   A technician attempted to sterilise a sample of cotton in a hermetically sealed glass bottle by autoclaving. Which of the following statements is correct?

   A   ❏ sterilisation cannot be achieved
   B   ❏ it should be sterilised at 45–118 °C for 30 min
   C   ❏ it should be sterilised at 121–124 °C for 15 min
   D   ❏ it should be autoclaved at 126–129 °C with saturated steam for 10 min
   E   ❏ it should be sterilised at 131–134 °C for 10 min

Q19   Sterilisation by radiation is achieved through the use of:

   1   ❏ ultraviolet (UV) radiation
   2   ❏ infrared radiation
   3   ❏ radiowaves
   4   ❏ gamma radiation
        A ❏ 1
        B ❏ 1 and 2
        C ❏ 1 and 3
        D ❏ 1 and 4
        E ❏ 1, 3 and 4

Q20   Evaluation of disinfectants is carried out by the:

   1   ❏ Rideal–Walker test
   2   ❏ Chick–Martin test
   3   ❏ Mantoux test
        A ❏ 1
        B ❏ 1 and 2
        C ❏ 1 and 3
        D ❏ 2
        E ❏ 1, 2 and 3

Q21   The Schick test is used for the diagnosis of:

   A   ❏ diphtheria
   B   ❏ lymphogranuloma
   C   ❏ syphilis
   D   ❏ leprosy
   E   ❏ tuberculosis

Questions 22 and 23 involve the following case:

A batch of parenteral formulation of ofloxacin for intravenous administration was prepared. The formulation was sterilised by normal autoclaving cycle and it passed the clarity and sterility tests. However, upon administration of the formulation, the patients showed signs of fever.

Q22   The likely cause of fever could be:

A   ❏ Gram-negative bacteria
B   ❏ Gram-positive bacteria
C   ❏ pyrogens
D   ❏ virus
E   ❏ chlamydiae

Q23   The animal model used in the official biological test to detect the cause of fever is:

A   ❏ mice
B   ❏ rabbits
C   ❏ rats
D   ❏ guinea pigs
E   ❏ beagle dogs

Questions 24 and 25 involve the following case:

A pharmacist dispensed an oral divided dose powder to a patient. After consuming the powder, the patient developed food poisoning.

Q24   Food poisoning is caused by spores of:

A   ❏ *Salmonella typhi*
B   ❏ *Staphylococcus albus*
C   ❏ *Pseudomonas aeruginosa*
D   ❏ *Clostridium welchii*
E   ❏ *Staphylococcus aureus*

Q25   Spores can be killed by:

A   ❏ 2% alcohol
B   ❏ 15% alcohol
C   ❏ boiling for 15 min

D ❑ heating for 2 min at 60 °C

E ❑ 2% glutaraldehyde

Q26 Which of the following is not required for administration of liquid medicines to children?

A ❑ a calibrated measure

B ❑ accuracy of measurement

C ❑ accuracy in delivery

D ❑ a paediatric dosing device

E ❑ a dose administration aid

Q27 Extemporaneous compounding is:

A ❑ a product for external use made by a pharmacist

B ❑ a medicine manufactured according to good manufacturing practice (GMP) principles

C ❑ the reformulation of a dosage form for a patient

D ❑ the provision of a liquid medicine for a patient

E ❑ the compounding of an unlicensed medicine

Q28 Licensing of medicines:

A ❑ was introduced in response to drug toxicity that affected adults and children

B ❑ is an unnecessary cost and process

C ❑ is not related to marketing authorisation, which includes dosage form, dose, patient age and route of administration

D ❑ is governed by international bodies such as the World Health Organization (WHO)

E ❑ is not relevant in extemporaneous compounding

Q29 'Therapeutic orphans' are paediatric patients who:

A ❑ do not have access to medicines

B ❑ have lost both parents

C ❑ use unlicensed medicines

D ❑ are treated by the government

E ❑ do not have access to medical care

Q30 The risks of extemporaneous compounding include:

A ❑ the potential for calculation errors

B ❑ selecting incorrect ingredients

C ❑ assignment of arbitrary shelf life

D ❏ incorrect quantities
E ❏ all of the above

Q31 Which of the following statements is false regarding the shelf life of compounded liquid products?

A ❏ it is necessary to use preservatives in all products
B ❏ they usually have few or no supporting stability data
C ❏ the expiry date is usually assigned arbitrarily
D ❏ compound hydroxybenzoate solution is sometimes used as a preservative
E ❏ it is best described as a 'beyond use' date

Q32 Which of the following statements regarding stable extemporaneously compounded topical preparations is false?

A ❏ they may be liquid, semisolid or paste dosage forms
B ❏ they may have a reduced expiry date if dilution has occurred
C ❏ they should be stored in well-closed containers in order to reduce evaporation
D ❏ they include betamethasone valerate in aqueous cream
E ❏ they do not need a preservative for a hydrophobic ointment base

Q33 Which of the following statements is false? Oral suspensions compounded from solid dosage forms:

A ❏ may include the use of compound tragacanth powder
B ❏ settle more slowly if a starch mucilage is incorporated
C ❏ redisperse more easily if tragacanth is incorporated
D ❏ may cake or result in uneven dosing without a suspending agent
E ❏ are flocculated by the inclusion of bentonite

Q34 An excipient that is less commonly used in compounded oral paediatric preparation is:

A ❏ glycerol for its sweet taste
B ❏ glycerol for increasing viscosity
C ❏ propylene glycol for increasing solubility
D ❏ ethanol for increasing solubility
E ❏ glycerol for increasing solubility

Q35 When compounding an oral product from a parenteral dosage form:

A ❑ the rate of absorption from the stomach may be increased
B ❑ the action of gastric acid may increase absorption and bioavailability
C ❑ the reformulation may result in precipitation
D ❑ the cost of reformulating a parenteral product is not an issue
E ❑ the bioavailability is always favourable

Q36 Palatability is one of the considerations when compounding oral liquid dosage forms. Which of the following statements is false?

A ❑ taste is more noticeable from soluble forms of the drug, due to diffusion to the taste buds
B ❑ insoluble complexes in suspensions reduce the bitter taste of drugs
C ❑ flavouring agents improve texture, taste and aroma
D ❑ flavouring agents eliminate the need for sweeteners
E ❑ lemon is a preferred flavour for some children

Q37 Taste-masking of drugs:

A ❑ is the use of saccharin and sodium cyclamate sweeteners
B ❑ is seldom required for suspensions of antibiotics
C ❑ can be achieved by using vanilla flavour
D ❑ involves objective assessment and the association with colour
E ❑ is normally required because of the association between taste and compliance

Q38 Which of the following is not used in pharmaceutical preparations as a sweetening agent?

A ❑ sorbitol
B ❑ mannitol
C ❑ sucrose
D ❑ lactose
E ❑ glycerol

Q39 Which of the following excipients would not be included for its effect on stability during the compounding of a liquid from crushed furosemide tablets?

A ❑ compound tragacanth powder
B ❑ methyl hydroxybenzoate
C ❑ xanthan gum

D ❏ propyl hydroxybenzoate
E ❏ sodium saccharin

Q40 Many community and some hospital pharmacies use a non-electronic beam balance for weighing ingredients. A beam balance consists of two pans, a set of weights, and an indicator that shows when the measured amount of ingredient in the right pan is equivalent to the mass of the weights in the left pan. Which of the following statement is false?

A ❏ weights should be cleaned regularly with metal cleaner in order to avoid rust
B ❏ strong draughts can affect the accuracy of weighing
C ❏ regular maintenance and revalidating is required for pharmaceutical scales
D ❏ the balance must be set to zero if weighing papers for greasy substances are used
E ❏ each scale has a minimum mass that can be weighed

Q41 Which of the following is not commonly used for compounding?

A ❏ conical measure
B ❏ mortar and pestle
C ❏ glass slab
D ❏ spatula
E ❏ volumetric flask

Q42 Select the true statement from the following about measuring small volumes of liquid during compounding:

A ❏ the contents of pipettes should be 'blown out' to maintain accuracy
B ❏ lids of containers should be placed upside-down on the bench-top during pouring
C ❏ the smallest cylinder that can be used to accurately measure 5 mL is a 50 mL cylinder
D ❏ measuring by difference is recommended for viscous liquids
E ❏ the smallest container that can be used to accurately measure 2 mL is a 25 mL cylinder

Q43 To perform a trituration for 20 mg of substance X (solubility 1 in 200 parts of water) using a scale that has a minimum recommended weight of 0.1 g, you should:

A ❏ weigh 20 g of substance X and add to the compounded mixture

B ❑ use a minimum of 20 mL of water for dissolving substance X

C ❑ use one-quarter of the volume of solution X for compounding

D ❑ dissolve 100 mg of substance X in 2 mL of water for compounding

E ❑ add 20 mg of substance X to the compounded mixture

Q44 Given the following formula for sodium bicarbonate ear drops BP, select the correct option:

a Sodium bicarbonate 500 mg (solubility 1 g in 11 parts water)

b Glycerol 3 mL

c Freshly boiled and cooled water to 10 mL

A ❑ to make 50 mL of ear drops, weigh 25 g of sodium bicarbonate

B ❑ to make 10 ml of ear drops, dissolve sodium bicarbonate in 55 mL of water

C ❑ 3 mL glycerol should be measured in a 15 mL conical measure

D ❑ dissolve 2.5 g of sodium bicarbonate in 30 mL of water, add 15 mL glycerol and make up the volume with water

E ❑ 3 mL glycerol should be measured in a pipette

Q45 Which of the following statements about chloroform is false?

A ❑ chloroform has a preservative action

B ❑ chloroform is volatile

C ❑ chloroform is carcinogenic

D ❑ chloroform has proved fatal in compounded medicines

E ❑ chloroform water double strength and chloroform water concentrate are interchangeable

Q46 Identify the false statement from the following:

A ❑ emulsions can be compounded from the contents of vitamin D capsules

B ❑ emulsions are thermodynamically unstable and should be refrigerated

C ❑ oral liquids of vitamins A, D, E and K can be toxic in large quantities

D ❑ to form emulsions, acacia can be used in fixed ratio combinations with oil and water

E ❑ emulsions require less preservative due to the oil content of the mixture

Q47 Select the false statement from the following:

A ❏ containers should preserve the quality of a medicine for the expected shelf life

B ❏ glass is the preferred material for packaging compounded liquids

C ❏ fluted bottles are used for paediatric compounded liquids

D ❏ plastic containers may leach ingredients

E ❏ compounded creams are often packed in wide-mouthed jars

Q48 Select the incorrect statement from the following:

A ❏ a mortar holds the ingredients while the pestle is used for grinding

B ❏ porcelain mortars and pestles are used for compounding emulsions

C ❏ the size of the mortar and pestle must be considered when compounding emulsions

D ❏ cytotoxic tablets should be reformulated on the bench-top using a glass mortar and pestle

E ❏ a glass mortar and pestle are used for grinding tablets for reformulation

Q49 Select the false statement from the following:

A ❏ paints are liquid products that are applied to the skin

B ❏ alcohol is a solvent used in paints

C ❏ collodions leave a tough, flexible film that is protective

D ❏ castor oil can be used in collodions to increase the flexibility

E ❏ alcohol is avoided because of its volatility

Q50 Select the true statement from the following:

A ❏ elixirs are compounded by diluting cough mixtures with ethanol

B ❏ eye preparations are compounded under aseptic conditions

C ❏ irrigations are compounded extemporaneously in hospitals

D ❏ compounded ear preparations have to be sterile

E ❏ oil is added while compounding nasal drops to lubricate the nasal passage

Q51 Select the true statement from the following:

A ❏ mint flavour is associated with liquid products for indigestion

B ❏ mint has a taste but not a fragrance

C ❑ each synthetic flavour has a natural counterpart
D ❑ aromatic oils cannot be formulated as alcoholic extracts
E ❑ taste buds on the tongue, cheeks and throat interact with suspended particles

Questions 52–55 involve the following case:

The pharmacist receives a prescription for 100 mL of trimethoprim mixture (10 mg/mL) with a prescribed dose of 4 mL *nocte*. The patient is 4 years old and weighs 16 kg.

Q52 If the recommended dose range is 2–4 mg/kg, the dose prescribed is:

A ❑ 2.5 mg/kg
B ❑ 32 mg/kg
C ❑ 40 mg/kg
D ❑ an underdose
E ❑ an overdose

Q53 Trimethoprim is available in tablets of 300 mg strength. What is the minimum number of tablets required to be crushed in order to accurately compound the mixture?

A ❑ 3
B ❑ 3.33
C ❑ 4
D ❑ 5
E ❑ 300

Q54 The formula for the mixture is:
a Trimethoprim 1 g (or equivalent number of tablets, each containing 300 mg drug)
b Syrup to 100 mL
Once the tablets have been crushed and mixed with some syrup, what is the volume (mL) of compounded mixture required to produce the trimethoprim 10 mg/mL?

A ❑ 100
B ❑ 110
C ❑ 120
D ❑ 150
E ❑ 900

Q55 If 150 mL of trimethoprim mixture was prescribed and the doctor also added compound methylhydroxybenzoate solution at a concentration of 1% v/v as a preservative, the volume (in mL) of solution required is:

A ❑ 0.5
B ❑ 1.0
C ❑ 1.5
D ❑ 2.0
E ❑ 2.5

Q56 Select the correct option from the following. Medicines must meet acceptable standards of:

A ❑ safety
B ❑ quality
C ❑ efficacy
D ❑ safety and efficacy
E ❑ quality, safety and efficacy

Q57 ICH stands for

A ❑ International Childhood Federation
B ❑ International Conference on Harmonisation
C ❑ International Convention on Harmonisation
D ❑ International Conference on Harmony
E ❑ International Clinical Harmonisation

Q58 Which of the following is not a benefit of ICH-directed harmonisation?

A ❑ a reduction in the number of tests that must be repeated using slightly different conditions in order to meet different countries' requirements
B ❑ reduced cost of research and development in the pharmaceutical sector
C ❑ no need for inspection of pharmaceutical facilities
D ❑ faster registration times for new medicines
E ❑ consistency in registration documentation in various countries

Q59 The most appropriate definition of quality with reference to pharmaceuticals is:

A ❑ excellence
B ❑ doing one's best
C ❑ good performance

D   ❏ conformance to requirements

E   ❏ best in the lot

Q60   Which of the following statements is true?

A   ❏ it is possible to eliminate errors completely

B   ❏ prevention of errors is usually costlier than fixing the implications of errors

C   ❏ the quality of pharmaceuticals normally cannot be judged by patients or even by dispensing pharmacists

D   ❏ quality issues with pharmaceuticals are of no greater significance than quality issues with other commodities such as mobile phones

E   ❏ regulation of pharmaceuticals in not related to quality – manufacturers do a good job of making good-quality products anyway

Q61   In the case of pharmaceuticals, quality is the responsibility of:

A   ❏ the management of the manufacturing organisation

B   ❏ quality control

C   ❏ the production supervisor and supply controller

D   ❏ the dispensing pharmacist

E   ❏ all of the above

Q62   GXPs that are most relevant to pharmaceuticals include:

A   ❏ GLP, GMP and GAP

B   ❏ GLP, GMP and GCP

C   ❏ GLP, GMP and GDP

D   ❏ GMP, GDP and GCP

E   ❏ GLP, GDP and GAP

Q63   Which of the following are examples of audit techniques?

A   ❏ trace forward

B   ❏ trace backward

C   ❏ random

D   ❏ trace all sides

E   ❏ A, B and C

Q64   Which of the following are potential sources of contamination in pharmaceutical manufacturing operations?

1   ❏ operators

2   ❏ dust and microbes from the atmosphere

3   ❏ active ingredient cross-contamination

A ❏ 1
B ❏ 2
C ❏ 3
D ❏ 1 and 2
E ❏ 1, 2 and 3

Q65   Specifications in pharmaceutical quality control:

1   ❏ serve as a quality-control tool
2   ❏ set the standards for materials and quality
3   ❏ describe tests to be carried out to comply with the selected
standards

A ❏ 1
B ❏ 2
C ❏ 1, 2 and 3
D ❏ 3
E ❏ 1 and 2

Q66   Which of the following is unsuitable for archiving data?

A   ❏ paper records written in ink
B   ❏ magnetic tapes
C   ❏ thermographic printouts
D   ❏ data on CD-ROM
E   ❏ data on servers

Q67   Which of the following would not normally be archived at the end of a
good laboratory practice (GLP) study?

A   ❏ creams
B   ❏ tablets
C   ❏ suspensions
D   ❏ plasma samples
E   ❏ active drug powder

Q68   All equipment that generates data goes through a qualification process before being used in a good laboratory practices (GLP) study.
The correct sequence for qualification of equipment, where DQ is
design qualification, IQ is installation qualification, OQ is opera-
tional qualification and PQ is performance qualification, is:

A   ❏ IQ > PQ > OQ > DQ
B   ❏ DQ > IQ > OQ > PQ
C   ❏ DQ > IQ > PQ > OQ

D ❏ PQ > OQ > IQ > DQ
E ❏ IQ > OQ > PQ > DQ

Q69 During quality control testing, a sample did not meet specification, i.e. an out-of-specification (OOS) result was obtained. The correct option is to:

A ❏ retest until the sample meets specification
B ❏ retest, but with a new sample aliquot
C ❏ retest only after an OOS is reported and investigated
D ❏ accept the test result as final and discard the sample
E ❏ ignore the result and clear the sample

Q70 Which of the following statements is incorrect?

A ❏ quality assurance (QA) is the sum total of organised arrangements made with the object of ensuring that medicinal products are of the quality required for their intended use
B ❏ quality control (QC) is that part of good manufacturing practices (GMP) concerned with sampling, specification and testing, and with the organisational, documentation and release procedures that ensure that the necessary and relevant tests are actually carried out and that the materials are not released for use, nor products released for sale or supply, until their quality has been judged to be satisfactory
C ❏ quality assurance (QA) is a part of QC
D ❏ QC involves determining the acceptability of each batch for release
E ❏ good laboratory practices (GLP) are a set of principles and procedures that, when followed by laboratory studies, help ensure that the data generated can be used to assess hazards and risks

Q71 CFR 211 refers to:

A ❏ Code of Federal Regulations 211
B ❏ Compliance to Federal Regulations 211
C ❏ Code of Final Regulations 211
D ❏ Code of Federal Requirements 211
E ❏ Compliance to Federal Requirements 211

Q72 As per good manufacturing practices (GMP), it is mandatory to have written specifications for:

1 ❑ active ingredients
2 ❑ excipients
3 ❑ packaging materials
A ❑ 1
B ❑ 2
C ❑ 1, 2 and 3
D ❑ 3
E ❑ 1 and 2

Q73 Which of the following will be considered a 'critical defect' during inspection of a pharmaceutical facility?

A ❑ lack of standard operating procedure (SOP) for equipment cleaning
B ❑ cracks in wall surfaces
C ❑ cross-contamination
D ❑ improper gowning
E ❑ correction in data entry done in such a way that the original entry is clearly visible

Q74 Which of the following statements is not true?

A ❑ standard operating procedures (SOPs) contain step-by-step instructions for carrying out a specific task
B ❑ batch manufacturing and packaging instructions refer to one or more SOPs
C ❑ preparation, issuing and updating of SOPs are controlled by analysts
D ❑ pharmacists are responsible for ensuring quality of products
E ❑ photocopying of SOPs is permitted, provided that written protocols are followed

Q75 Which two countries from the Organisation for Economic Co-operation and Development (OECD) group permit direct-to-consumer advertising (DTCA) of pharmaceutical products?

A ❑ the USA and the UK
B ❑ New Zealand and the USA
C ❑ the UK and New Zealand
D ❑ Australia and New Zealand
E ❑ Singapore and Australia

238 | MCQs in Pharmaceutical Science and Technology

Q76 According to the ICH guideline Q10, i.e. Pharmaceutical Quality System, which of the following is not part of the product lifecycle of a drug substance or drug product?

A ❏ pharmaceutical development
B ❏ technology transfer
C ❏ manufacturing
D ❏ product recalls
E ❏ distribution

Q77 Which of the following statements regarding the ICH is false?

A ❏ ICH is a tripartite agreement between Europe, Japan and the USA
B ❏ ICH guidelines are legally binding and require US Food and Drug Administration (FDA) approval
C ❏ ICH guidelines are designed to help reduce differences in requirements between different countries, allowing for faster introduction of medicines and increased availability to patients
D ❏ ICH guidelines help coordinate technical requirements, leading to greater mutual acceptance of research
E ❏ ICH membership includes industry associations from each of the three tripartite areas

Q78 Post-marketing surveillance of phase IV trials are conducted after a drug has received approval for marketing. The objectives of phase IV trials is (are) to:

1 ❏ find more information regarding the safety and side-effect profile of a drug
2 ❏ evaluate the long-term risks and benefits of a drug
3 ❏ gather data from a larger population base than was used during clinical trials

A ❏ 1
B ❏ 2
C ❏ 1, 2 and 3
D ❏ 3
E ❏ 1 and 2

Q79 Which of the following is not the responsibility of a drug regulatory affairs team?

1 ❏ interaction with regulatory bodies
2 ❏ dossier compilation, publishing, printing and archiving

3   ❑ setting the launch date of a new product
A ❑ 1
B ❑ 2
C ❑ 1, 2 and 3
D ❑ 3
E ❑ 1 and 2

Q80 Each stage of clinical trials is designed to gather specific information. Which of the following statements regarding phase I clinical trials is false?

A   ❑ the study is conducted in healthy volunteers in most cases
B   ❑ the number of participants is usually small, in the range 6–12
C   ❑ pharmacokinetic and pharmacodynamic data are gathered
D   ❑ the human tolerability of new drug molecules is assessed
E   ❑ the efficacy of drugs is evaluated, which acts as a prequalifying requirement for the next phase, i.e. phase II

Q81 GCP stands for:

A   ❑ good clinical practice
B   ❑ good customer practice
C   ❑ good clinical policy
D   ❑ good consumer policy
E   ❑ good consumer practice

Q82 Which of the following refers to essential document(s) required before a clinical trial commences?

A   ❑ investigator brochure
B   ❑ informed consent form
C   ❑ documented approval from the ethics committee
D   ❑ signed agreements by all parties
E   ❑ all of the above

Q83 Which of the following is classified as fraud in clinical studies?

A   ❑ deliberately reporting false or misleading data
B   ❑ altered data
C   ❑ fabricated data
D   ❑ data from observations that were not documented
E   ❑ all of the above

Q84 As per the Declaration of Helsinki, which of the following statements is not true?

A ❑ the clinical studies are to be scientifically validated and qualified

B ❑ volunteers are to be well informed; however, they cannot withdraw after trials have started

C ❑ placebo medication should be used only in the absence of existing proven therapy

D ❑ the study should be based on scientific principles

E ❑ information about the potential benefits, risks, burdens and effectiveness is shared with volunteers

Q85 As per the Declaration of Helsinki, which of the following rights is (are) recommended to safeguard research participants?

A ❑ the right not to be harmed

B ❑ the right to full disclosure

C ❑ the right to self-determination

D ❑ the right of privacy

E ❑ all of the above

Q86 As per ICH guidelines on clinical trials, the trial investigators:

1 ❑ must be appropriately qualified

2 ❑ must have the resources and time to conduct the trial

3 ❑ must permit monitoring, auditing and inspections

A ❑ 1

B ❑ 2

C ❑ 1, 2 and 3

D ❑ 1 and 3

E ❑ 1 and 2

Q87 As per ICH guidelines on clinical trials, trial participants are entitled to:

1 ❑ a qualified investigator who is responsible for trial-related decisions

2 ❑ adequate care during trial

3 ❑ freedom to withdraw consent, rights of respect, confidentiality, disclosure and self-determination

A ❑ 1

B ❑ 2

C ❑ 1, 2 and 3

D ❑ 1 and 3
E ❑ 1 and 2

Q88 As per ICH guidelines on clinical trials, informed consent of the trial participants is mandatory. The consent describes elements of:

A ❑ risks
B ❑ benefits
C ❑ privacy
D ❑ cultural
E ❑ all of the above

Q89 Which of the following statements is true with reference to informed consent in a clinical trial?

A ❑ anybody can inform the patient, as long as the information provided is complete
B ❑ there is no requirement that a medically qualified person has to obtain the consent
C ❑ the consent can be obtained any time during the first week of the clinical trial
D ❑ the documentation and procedure have to be the same in all countries in the case of multi-country trials
E ❑ the consent does not necessarily have to be in a language that the trial participants can understand

Q90 Which one of the following is not a red flag for fraud in a clinical data set?

A ❑ exact number of days between visits
B ❑ perfect tablet counts
C ❑ poor filing of documents
D ❑ no dropouts
E ❑ no or few adverse events

Q91 In pharmaceutical dosage forms, packaging provides which of the following:

A ❑ identification
B ❑ protection
C ❑ compliance
D ❑ convenience
E ❑ all of the above

Q92 Which of the following is not true with reference to packaging?

A ❑ good packaging can protect a thermolabile drug for a long time

B ❑ packaging requirements in different parts of the world can be different for the same product, e.g. protection from temperature extremes that the product is likely to experience during its shipment and shelf life

C ❑ depending upon the moisture levels inside and outside a product, water can permeate in or out of the packaging

D ❑ photodegradation can be avoided through the use of coloured opaque containers

E ❑ fragrances and flavouring agents can permeate through packaging, causing a change in colour or taste

Q93 During transportation and storage, pharmaceutical products need mechanical protection from which of the following:

1 ❑ compression
2 ❑ impact
3 ❑ vibration

A ❑ 1
B ❑ 2
C ❑ 3
D ❑ 1 and 2
E ❑ 1, 2 and 3

Q94 Which of the following statements is not true with reference to biological hazards to pharmaceutical products?

A ❑ closure is an important component for microbial protection

B ❑ products for repeated use should be adequately preserved

C ❑ contamination of products from insects or rodents is not an issue

D ❑ sterile products should be capable of handling the sterilisation process

E ❑ microorganisms can attack the packaging itself

Q95 Which of the following factors are to be considered when selecting an appropriate packaging?

1 ❑ product composition and stability profile
2 ❑ regulatory requirements

3 ❑ cost
   A ❑ 1
   B ❑ 2
   C ❑ 3
   D ❑ 1 and 2
   E ❑ 1, 2 and 3

Q96 Which one of the following is not a common packaging option for solid dosage forms?

A ❑ bottles
B ❑ blisters
C ❑ foils
D ❑ flexible tubes
E ❑ metallic and lined containers

Q97 Which of the following statements is not true for glass used in the pharmaceutical industry?

A ❑ type I, known as borosilicate type, is the least reactive
B ❑ type II, known as treated soda lime glass, is acceptable for most products except blood products
C ❑ type III has average hydrolytic resistance and is suitable for non-aqueous preparations
D ❑ type IV glass is the least resistant to hydrolytic attack and is suitable for solid products
E ❑ type V is the most resistant glass type and can be used for anything

Q98 Which of the following statements is not true with reference to poly-ethylene and polypropylene as the packaging materials?

A ❑ higher-density grades provide greater rigidity and melting point
B ❑ low-density grades are more flexible
C ❑ polypropylene is widely used to make closures
D ❑ all grades of polyethylene and polypropylene are resistant to stress cracking
E ❑ high-density polyethylene is used for bottles and has good moisture resistance

Q99   Which one of the following is not a true statement regarding packaging?

A   ❏ A number of plastics, including PVC and LDPE, can be used to make films

B   ❏ LDPE is the most suitable polymeric material for heat sealing

C   ❏ the moisture barrier of HDPE film is good when compared with other plastic films

D   ❏ the inner layer of laminates is polyethylene, which is non-reactive, and therefore compatibility studies are not required

E   ❏ a sachet can have as many as four layers, including layers for information, mechanical protection and moisture resistance, and a heat-sealable layer

Q100   Select the correct option from the following with reference to regulatory quality control of packaging materials:

1   ❏ as per good manufacturing practices (GMP), packaging materials are treated similarly to active ingredients and excipients for their quality, sampling and documentation

2   ❏ pharmacopoeias do not cover packaging materials

3   ❏ compatibility testing of a pharmaceutical product and packaging material is an integral part of the product development programme

A ❏ 1
B ❏ 2
C ❏ 3
D ❏ 1 and 3
E ❏ 1, 2 and 3

Answers

A1   A

Biological indicators (BIs) consist of standardised bacterial spore preparations. These are used primarily for validation rather than routine monitoring
of thermal, chemical or radiation sterilisation processes. Detailed use is shown
in Table 6.1.

A2   C

Bacteria are divided into two groups, designated Gram-positive and Gram-
negative according to their reaction to a staining procedure developed in
1884 by Christian Gram. Gram-positive organisms are able to retain the
crystal violet stain because of the high amount of peptidoglycan in the cell
wall. Gram-positive cell walls typically lack the outer membrane found in
Gram-negative bacteria.

A3   B

Polio vaccine is given by both the oral and parenteral routes. Oral polio
vaccine (OPV) is a live-attenuated vaccine and preferred over the injectable
inactivated polio vaccine (IPV) as it eliminates the need for sterile syringes and
makes the vaccine more suitable for mass vaccination campaigns. OPV also
provides longer-lasting immunity than IPV.

A4   E

The primary immune response is a relatively weak immune response that
occurs when naive lymphocytes first encounter an antigen. IgM is essentially

Table 6.1 Sterilisation methods and biological indicators	
Sterilisation method	**Biological indicators**
Moist heat	Spores of *Bacillus stearothermophilus*, *B. subtilis*, *B. coagulans* and *Clostridium sporogenes*
Dry heat	Spores of *B. subtilis*
Ionising radiation	Spores of *B. pumilus*
Filtration	*Pseudomonas diminuta*
Ethylene oxide	Spores of *B. subtilis* var. niger
Hydrogen peroxide	Spores of *B. stearothermophilus*, *B. subtilis*, *B. coagulans* and *C. sporogenes*

intravascular and is produced early in the immune response. Because of its high valency, it is a very effective bacterial agglutinator and mediator of complement-dependent cytolysis and is therefore a powerful first-line defence against bacteraemia.

A5   A

HEPA filters are filters composed of glass fibres and fillers and considered to be at least 99.97% efficient in removing particles of 0.3 μm size and larger. They are used mainly to remove finer debris in the submicrometre range, including microorganisms from intensive care units (ICU) of hospitals and sterile zones in laboratories.

A6   D

Dengue fever is an acute febrile disease found in the tropics. Dengue is transmitted to humans by *Aedes aegypti* or more rarely *A. albopictus* mosquitoes, which feed during the day.

A7   B

Bacillus Calmette-Guérin (or Bacille Calmette-Guérin, BCG) is a bacterial vaccine against tuberculosis that is prepared from a strain of the attenuated (weakened) live bovine tuberculosis bacillus, *Mycobacterium bovis*, which has lost its virulence in humans by being specially cultured in an artificial medium for years.

A8   A

An important term in expressing microbial death kinetics for heat, chemical and radiation sterilisation is the D value. The D value is the time (for heat or chemical exposure) or the dose (for radiation exposure) required for the microbial population to decline by one decimal point (a 90%, or one logarithmic unit, reduction).

A9   C

Ethylene oxide is an alkylating gas used for sterilisation of medical devices. It alkylates essential metabolites of microorganisms, particularly affecting the reproductive process. The altered metabolites are not available for reproduction, resulting in the death of the microorganisms.

A10 E

Typhus is a diseases caused by *Rickettsia prowazekii*. The name comes from the Greek *typhos*, meaning smoky or hazy, describing the state of mind of people affected with typhus. The causative organism *Rickettsia* is an obligate parasite and cannot survive for long outside living cells. Typhus should not be confused with typhoid fever – the two are unrelated.

A11 D

Q fever is caused by infection with *Coxiella burnetii*, a bacterium that affects both humans and animals. The infection results from inhalation of contaminated particles in the air, and from contact with the milk, urine, faeces, vaginal mucus or semen of infected animals. The bacterium is an obligate intracellular pathogen and is uncommon, but it may be found in cattle, sheep, goats and other domestic mammals, including cats and dogs.

A12 A

The structure of the herpes virus consists of a relatively large double-stranded, linear DNA genome encased within an icosahedral protein cage called the capsid, which is wrapped in a lipid bilayer called the envelope.

A13 C

DTP (also DTwP and DPT) refers to a combination of vaccines against three infectious diseases in humans: diphtheria, tetanus and pertussis (whooping cough). The vaccine components include diphtheria and tetanus toxoids, and killed whole cells of the organism that causes pertussis (wP).

A14 D

The Widal test is a presumptive serological test for undulant fever or enteric fever. In the case of *Salmonella* infections, it is a demonstration of agglutinating antibodies against antigens O-somatic and H-flagellar in the blood.

A15 B

Interferons (IFNs) are natural cell signalling proteins (e.g. glycoproteins known as cytokines) produced by the cells of the immune system of most vertebrates in response to the presence of double-stranded RNA, a key indicator of viral infections. Interferons assist the immune response by inhibiting viral replication within host cells, activating natural killer cells and macrophages, increasing antigen presentation to T-lymphocytes, and increasing the resistance of host cells to viral infection.

A16   E

Commercially, citric acid is produced through fungal (*Aspergillus niger*) fermentation. Chemical synthesis of citric acid is possible, but it is not cheaper than fungal fermentation. Usually *A. niger* is chosen over other potential citrate-producing organisms because it uses cheap raw materials (molasses) as substrate and produces a high consistent yield.

A17   B

Clean rooms are classified into various classes depending on their cleanliness. For example, a class 100 000 room is one in which the particle count is no more than 100 000 per cubic foot of 0.5 μm and larger in size. Usually class 10 000 aseptic areas are used to store or transfer products or components that involve a substantial risk of environmental contamination. Class 100 areas are used where it is critical to maintain product sterility. Class 10 and class 1 areas are normally found in the microelectronics industry. All beta-lactam antibiotics, including penicillin, are prepared in class 10 clean rooms.

A18   A

Autoclaving involves sterilisation at 121–124 °C for 15 min. It is used to sterilise culture media, various surgical stainless-steel instruments, various types of surgical dressing, gauzes and sutures, solutions and syringes. In order to sterilise the surface of a solid, the 'steam' needs to come into contact with the surface. Since the sample of cotton that is to be sterilised is hermetically sealed in a glass bottle, sterilisation cannot be achieved, as steam will be unable to make contact with all the exposed surface of the cotton.

A19   D

UV and gamma radiations are used as terminal methods of sterilisation. UV light is commonly used to aid in the reduction of contamination in the air and on surfaces within the processing environment. When UV light passes through matter, energy is liberated to the orbital electrons within constituent atoms. This absorbed energy causes a highly energised state of the atoms and alters their reactivity. When such excitation and alteration of activity of essential atoms occur within the molecules of microorganisms or of their essential metabolites, the organism dies or is unable to reproduce. Gamma or ionising radiation destroys microorganisms by stopping reproduction as a result of lethal mutations. These, in turn, bring about energy changes in nucleic acids and other molecules, thus eliminating their availability for the metabolism of the bacterial cell. Ionising radiation differs from UV rays in its effects on matter, primarily in that the former is of a higher energy level and actually produces ionisation of constituent atoms.

A20   A

The Rideal–Walker test was developed by English chemists Samuel Rideal (1863–1929) and JT Ainslie Walker (1868–1930) to evaluate disinfectants. The Rideal–Walker coefficient is a figure that expresses the disinfecting power of any substance. It is obtained by dividing the figure linked to the degree of dilution of the disinfectant that kills a microorganism in a given time by the figure that indicates the degree of dilution of phenol that kills the organism in the same length of time under similar conditions.

A21   A

The Schick test is used to determine whether a person is susceptible to diphtheria. In this test, a small amount (0.1 mL) of diluted (1/50 MLD) diphtheria toxin is injected intradermally into the person's arm. If the person does not have enough antibodies to fight it off, the skin around the injection will become red and swollen, indicating a positive result. This swelling disappears after a few days. If the person has immunity, then little or no swelling and redness will occur, indicating a negative result.

A22   C

Pyrogens are products of metabolism of microorganisms. Gram-negative bacteria produce the most potent pyrogenic substances as endotoxins. Chemically, pyrogens are lipid substances associated with a carrier molecule, which is usually a polysaccharide but may be a peptide.

A23   B

The presence of pyrogenic substances in parenteral preparations is determined by a qualitative biologic test based on the fever response of rabbits (pyrogen test). Rabbits are used as the test animal because they show a physiologic response to pyrogens similar to that of human beings. If a pyrogenic substance is injected into the vein of a rabbit, an elevation of temperature occurs within a period of 3 h.

A24 D

Clostridium welchii type A is a common cause of food poisoning when it is allowed to proliferate to large numbers in cooked foods, usually meat and poultry. The main factors of importance are survival of the spores frequently found on raw products, through the cooking process, and possible contamination of cooked meats transferred to unclean containers. Subsequent germination of spores and rapid multiplication of the vegetative cells during long slow cooling or non-refrigerated storage leads to heavy contamination. The toxin responsible is different from the soluble antigens, and its formation in the intestine is associated with sporulation.

A25 E

In vitro exposure to 2% alkaline glutaraldehyde for 5 min will result in 99.0% or greater killing of spores.

A26 E

A dose administration aid is the term used to describe blister packaging to aid compliance.

A27 C

Extemporaneous compounding is the provision of a medicine in a suitable dosage form for a particular patient. A medicine may be unlicensed in a particular country for a variety of reasons.

A28 E

The licensing of medicines includes not only the clinical data and indications but also the age group and specific doses for specific conditions.

A29 C

Many medicines for children are often not tested in that age group, and hence doctors use them by 'trial and error'.

A30 E

Extemporaneous compounding is associated with risks of calculation errors, incorrect ingredients, arbitrary shelf life and formulation errors.

A31 A

Compounded products without a preservative will have a short 'expiry' date; some patients are allergic to preservatives and hence they are specifically compounded extemporaneously.

A32 D

Betamethasone valerate has a cationic charge and aqueous cream is anionic. Charge interaction results in changes to the betamethasone molecule over time, resulting in a less effective product.

A33 E

Bentonite is a suspending agent used in the compounding of external preparations.

A34 D

The use of ethanol has decreased in recent times, and potentially less toxic excipients or combinations of excipients are preferred for children.

A35 C

Parenteral products often contain excipients to aid solubility. If these are diluted during reformulation, precipitation can occur.

A36 D

A flavour is not necessarily sweet. Once a flavour has been selected, reformulation may include the addition of syrup or artificial sweeteners. A limit in the use of saccharin has been recommended by the American Medical Association.

A37 E

Adding a sweetener is not sufficient for taste-masking. Taste is a subjective assessment and influences compliance, especially in paediatric patients.

A38 D

The relative sweetness of lactose is low (only 20%) when compared with sucrose (100%).

A39   E

Sodium saccharin is a sweetener. The other ingredients are a suspending agent (compound tragacanth powder) or agents to increase viscosity (xanthan gum) and reduce sedimentation rate, and preservatives.

A40   A

Weights should not be touched with fingers or substances, in order to maintain their accuracy.

A41   E

A conical flask is used to accurately prepare a solution such as a standard for use on a UV-spectrophotometer. It may be used in compounding, but it is very unlikely.

A42   D

Accurate measurement of a liquid should be done on a 1 : 5 basis, e.g. 2 mL in a 10 mL cylinder or 5 mL in a 25 mL cylinder. Viscous liquids need to be measured by difference because much of the liquid stays behind in the measuring cylinder after pouring.

A43   B

20 mg needs to be multiplied by a factor of 5 to weigh 100 mg minimum. This will require (0.1 g × 200) parts of water for minimum solubility. Weigh 100 mg, dissolve in 20 mL, and add one-fifth of the solution (i.e. 4 mL) to the compounded mixture in order to get 20 mg of substance X.

A44   D

Glycerol is too viscous for a pipette or a small measure: add by difference. The weights and volume selected should take into account the solubility.

A45   E

The double strength solution contains less chloroform than the concentrate.

A46   E

Preservatives can partition into oil, and therefore higher concentrations are often needed to maintain the required concentration in the aqueous phase.

A47   C

Fluted bottles are used for products intended for external use.

A48   D

Cytotoxic tablets should not be crushed because of the potential toxicity that may be caused by inhalation of the chemicals, unless special precautions are taken to protect the compounding pharmacist.

A49   E

In paints, alcohol, acetone or ether are used as solvents because they evaporate quickly, leaving a film on the skin.

A50   B

Irrigations are sterile preparations compounded under aseptic conditions. Eye preparations, but not necessarily ear drops, have to be sterile. Oil droplets can cause pneumonia if inhaled.

A51   A

Molecules must be in solution to produce a positive or negative taste sensation. Oils are formulated as alcoholic extracts, which presents them as miscible with water. Selection of flavour is guided by the intended purpose; the mint flavour is well related to digestion products.

A52   A

4 mL of mixture contains 40 mg of active ingredient. 40 mg divided by 16 kg is 2.5 mg/kg.

A53   C

The total amount of drug required is 1000 mg, which can be obtained from 3.33 tablets; however, 3.33 tablets cannot be accurately measured, and therefore 4 tablets should be crushed.

A54   C

Four tablets, each containing 300 mg of trimethoprim mixed to 120 mL of syrup, provides 10 mg/mL.

A55   C

1.5 mL of a 1% solution of methylhydroxybenzoate will provide 1% concentration.

A56   E

All medicinal products including new drugs are expected to meet certified and acceptable standards of quality, safety and efficacy.

A57   B

ICH stands for International Conference on Harmonisation of Technical Requirements for Registration of Pharmaceuticals for Human Use. Details are available at www.ich.org.

A58   C

ICH has harmonised registration documentation to a great extent and helps in standardising specifications such as levels of impurities; however, it has not obviated the need for inspection of facilities.

A59   D

All of the listed options are generally used to define quality. In the case of pharmaceuticals, 'quality' refers to compliance to certain standards, e.g. standards suggested in pharmacopoeial monographs.

A60   C

Although the errors can never be eliminated completely, they can be minimised. Quality issues with pharmaceuticals are monitored and well regulated because they deal with people's lives and any compromise can have serious implications. A number of tests to assess the quality are destructive, e.g. tablets have to be crushed before being assayed. Therefore, conclusions are drawn about the lots based on results from samples. A patient or pharmacist cannot assess a formulation completely simply by looking at it.

A61   E

Everyone involved with a pharmaceutical product, from manufacturing to quality control, distribution and dispensing, is responsible for the quality of a pharmaceutical.

A62 B

Good laboratory practices (GLP), good manufacturing practices (GMP) and good clinical practices (GCP) are directly relevant to pharmaceuticals. Good agricultural practices (GAP) are relevant in the case of herbal medicines, while good distribution practices (GDP) are indirectly covered as part of GMP.

A63 E

Audit can be random, trace backward and trace forward. As the name suggests, it is the sequence of tracking events or processes that dictates the type of audit.

A64 E

Contamination in pharmaceutical products can occur from a number of sources, including machines, operators, cross-contamination, people and environment.

A65 C

Specification can be developed in-house or provided by official compendia such as pharmacopoeial monographs, and act as quality standards. These govern the tests to be carried out on active ingredients and formulations, and give guidance on interpretation of results.

A66 C

Thermographic printouts tend to fade over a period of time. The data and documents stored in archives should be readable and traceable for the specified period of time.

A67 D

Plasma samples and other biological tissues are not stored in archives, for obvious reasons of hazard to human health.

A68 B

The correct sequence is DQ > IQ > OQ > PQ. In most cases, maintenance qualification (MQ) is added during the shelf life of an equipment.

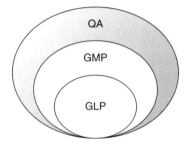

Figure 6.1 Scope of quality assurance (QA)

A69 C

Recording of OOS results followed by appropriate investigation is common practice in quality-control laboratories when unexpected results are obtained.

A70 C

Quality assurance is wider in its scope and covers GMP and GLP as per Figure 6.1.

A71 A

Code of Federal Regulations 211 refers to the US FDA GMP regulations. These are known by other names in different countries.

A72 C

The specifications for active ingredients, excipients, packaging materials and finished products act as limits. The quality of these components is assessed with reference to these specifications. The specifications can be from compendial sources or established in-house.

A73 C

Cross-contamination is a critical defect that can have serious implications.

A74 C

Preparation, issuing and updating of SOPs is controlled by quality assurance (QA), independently of analysts. Analysts are supposed to follow SOPs and can make recommendations to QA.

A75   B

New Zealand and the USA are the only two countries in the OECD group that permit DTCA. DTCA has its advantages and limitations, the most important advantage being better education of patients; the downsides are the risk for self-medication and the tendency of patients to influence doctors.

A76   D

Product recalls are not considered part of the product lifecycle, as recall does not necessarily happen with all products.

A77   B

ICH guidelines are developed as recommendations. Countries wishing to adapt them have to go through the acceptance process and make them part of their regulatory system independently.

A78   C

Phase IV trials provide data from population-wide use of a drug, typically under uncontrolled conditions.

A79   D

The launch date of a product is typically set by the marketing team. The regulatory affairs team has only to ensure that the necessary regulatory approvals are in place before that date.

A80   E

A phase I trial is primarily a safety study with focus on dose adjustments. The drug is administered to human beings for the first time, and therefore the study is conducted in healthy volunteers. Efficacy of the drug is typically evaluated in phase II and III clinical trials.

A81   A

GCP refers to good clinical practice, which is one of the three critical GXPs for pharmaceuticals. The other two are GMP and GLP.

A82   E

In addition to all of the listed documents, protocols and amendments, laboratory accreditation and normal ranges, calibration certificates and manuals of operations are also required.

A83   E

Observations that were not documented is not a sign of fraud: it simply means non-compliance to requirements.

A84   B

In any clinical trials, volunteers or trial participants have a right to withdraw at any time, without giving a reason. This right is usually protected and volunteers are informed about it.

A85   E

All of the regulations on clinical studies are designed to protect study participants. The rights to privacy, self-determination, full disclosure and not to be harmed are parts of that principle.

A86   C

The trial investigator must be appropriately qualified and trained in good clinical practices (GCP), must employ appropriately qualified staff, should have resources and time to conduct the trial, should be familiar with the protocol and investigational product, permit monitoring, auditing and inspections, and have the capability to recruit the agreed number of subjects.

A87   C

Trial participants have a number of rights, including the freedom to withdraw at any stage.

A88   E

In addition to the risks, benefits, privacy and cultural issues, the informed consent form also describes the compensation and flexibility to withdraw from the trial.

A89 B

There is no requirement in law that a medically qualified person obtains informed consent. However, delegated responsibilities must be documented in the 'authorised signatures and delegation log'.

A90 C

Poor filing of documents is a reflection of a lack of tidiness and a need for improvement in systems. It does not indicate the potential for fraud.

A91 E

Packaging of pharmaceutical products provides identification and label information, protection, compliance, convenience and specific applications, e.g. inhalers.

A92 A

Thermolabile drugs can be protected for a short time with packaging. Over a long period, packaging cannot maintain temperature, unless special casings are used.

A93 E

During shipment and storage, formulations are exposed to shocks, impact, vibrations and compression, with incidences such as dropping on the pharmacy floor.

A94 C

Pharmaceuticals are susceptible to attack by insects and rodents, similar to other consumer goods.

A95 E

A number of factors affect selection of packaging. These include compatibility with formulation, stability requirements, regulations, cost, aesthetics, convenience, special functions and flexibility of design.

A96 D

Flexible tubes are generally used for semisolids, not for solid dosage forms.

A97 E

According to the British Pharmacopoeia, there are only four types of glass that are used in pharmaceuticals. There is no type V glass.

A98 D

Higher-density grades of polypropylene and polyethylene are more resistant to stress cracking.

A99 D

Compatibility with the formulation is a critical requirement to be established with all packaging, including laminates.

A100 D

Pharmacopoeias provide monographs and specifications for a range of packaging materials. For example, the British Pharmacopoeia has specifications for polyethylene, polypropylene, plastic additives, silicone and a number of other packaging components.

Index

drug mass
    dosage, 113
    formulation, 112
drug measurement, body fluids, 51
drug solutions, concentrated aqueous, 8
drug suspensions, stability, 9–10
drug transport, energy, 43
drug-release mechanisms, prolonged release
    tablets, 153
dry binders, tablet manufacture, 151
dry heat sterilisation, biological indicators, 220
dry powder inhalers (DPIs), metered dose
    inhalers (MDIs), 174
drying techniques
    fluidised-bed drying, 104–105
    freeze drying, 104
    microwave drying, 105
    solute migration, 105
    spray drying, 104
DTCA. *see* direct-to-consumer advertising
duration of action, intravenous
    preparations, 190

electric sensing zone method (Coulter counter),
    particle size, 92
electrolytes
    parenteral preparations, 190
    solubility, 6
elimination half-life, digoxin, 52
elimination rate constant
    half-life, 51–52
    ibuprofen, 51–52
    intravenous injections, 58
emulgents, properties, 142
emulsification of creams, 144
emulsions, 12
    acacia, 143
    formulation, 141, 143
    intravenous injections, 141
    macromolecules, 11
    oil-in-water, 10
    oils, 141, 142
    particulate systems, 11
    physical stability, 142
    preservatives, 143
    properties, 10, 142, 230
    stability, 11, 17–18
    surfactants, 11, 17–18
    Tween 80, 17–18
enteric-coated systems, properties, 60
enterohepatic recycling, 43
equipment
    archiving data, 235–236
    good laboratory practices (GLP)
      study, 235–236
equipment/unit operation, tabletting, 150

erythromycin
    gastric acid, 5
    hydrolysis, acid-catalysed, 5
    stability, 5
eutectic mixtures, properties, 90
eutectic point, 90
excipients
    compatibility studies, 25
    creams, 143–144
    fillers, 150–151
    film coating, 108
    formulation development, 25
    furosemide tablets, 228–229
    oral paediatric preparations, 227
    stability, 228–229
    sugar coating of tablets, 107
    tablets, 150–151
expected maximum concentration, intravenous
    injections, 58, 59
expected minimum concentration, intravenous
    injections, 58, 59
extemporaneous compounding
    defining, 226
    risks, 226–227
extemporaneously compounded topical
    preparations, properties, 227
extrusion
    granulation, 103
    properties, 103
    solute migration, 103
    spheronisation, 105–106
eye preparations. *see* ophthalmic drugs

facilities, pharmaceutical, critical defects, 237
fentanyl, 177
    Fick's law of diffusion, 177
    stratum corneum penetration, 177
    transdermal delivery system, 177
fever
    animal model, 225
    ofloxacin, 225
Fick's law of diffusion
    fentanyl, 177
    rate of diffusion, 5–6
fillers
    excipients, 150–151
    tablets, 150–151
film coating
    excipients, 108
    tablet coating, 107, 108
filter aids, 137
filters
    size, 191
    validation, 138
filtration
    Darcy's equation, 137